Bioterrorism Preparedness

Edited by
Nancy Khardori

Related Titles

Jacquelyn G. Black

**Study Guide to accompany Mirobiology:
Principles and Explorations**
6th Edition

2005
ISBN 0-471-48244-7

Marc Siegel

False Alarm: The Truth about the Epidemic of Fear

2005
ISBN 0-471-67869-4

Richard F. Pilch, Raymond A. Zilinskas (Eds.)

Encyclopedia of Bioterrorism Defense

2005
ISBN 0-471-46717-0

Roberta Carroll, American Society for Healthcare Risk Management
(ASHRM) (Eds)

Risk Management Handbook for Health Care Organizations
4th Edition

2003
ISBN 0-7879-6797-1

Bioterrorism Preparedness

Medicine – Public Health – Policy

Edited by
Nancy Khardori

WILEY-VCH Verlag GmbH & Co. KGaA

The Editors

Dr. Nancy Khardori
Southern Illinois Universitiy
School of Medicine
Division of Infectious Diseases
701 North First Street
Springfield, Illinois 62794-9636
USA

■ All books published by Wiley-VCH are carefully produced. Nevertheless, authors, editors, and publisher
do not warrant the information contained in these
books, including this book, to be free of errors.
Readers are advised to keep in mind that statements,
data, illustrations, procedural details or other items
may inadvertently be inaccurate.

Library of Congress Card No.:
applied for

British Library Cataloguing-in-Publication Data
A catalogue record for this book is available from the
British Library.

**Bibliographic information published by
Die Deutsche Bibliothek**
Die Deutsche Bibliothek lists this publication in the
Deutsche Nationalbibliografie; detailed biliographic
data is available in the Internet at
<http://dnb.ddb.de>.

© 2006 WILEY-VCH Verlag GmbH & Co. KGaA,
Weinheim

Typesetting Dörr + Schiller GmbH, Stuttgart
Printing Strauss GmbH, Mörlenbach
Binding Litges & Dopf Buchbinderei GmbH,
Heppenheim
Cover Design 4T Matthes + Traut, Darmstadt

Printed in the Federal Republic of Germany
Printed on acid-free paper

ISBN-13: 978-3-527-31235-1
ISBN-10: 3-527-31235-8

List of Contents

1   **Potential Agents of Bioterrorism:**
         Historical Perspective and an Overview   *1*
         Nancy Khardori
1.1   Historical Perspective – How We Got Here   *1*
1.2   Development of Modern Biological Weapons   *3*
1.3   Biological Weapons Systems   *8*
1.4   Potential Bioterrorism Agents – Categorization and Prioritization   *10*
1.5   Category B – Bacterial/Rickettsial Agents of Bioterrorism   *14*
1.5.1   Brucellosis   *14*
1.5.2   Glanders and Melioidosis   *15*
1.5.3   Psittacosis   *17*
1.5.4   Q Fever   *18*
1.5.5   Typhus Fever   *19*
1.5.6   Food and Water Safety Threats   *21*

1.6   Category B – Viral Agents of Bioterrorism   *22*
1.6.1   Alphavirus Encephalomyelitis   *22*

1.7   Category B – Biological Toxins for Bioterrorism   *23*
1.7.1   Enterotoxin B   *23*
1.7.2   Epsilon (Alpha Toxin)   *24*
1.7.3   Ricin Toxin   *24*
1.7.4   T-2 Mycotoxins   *24*

1.8   Other Toxins With Potential for Bioterrorism   *25*
1.8.1   Nipah and Hendra Viruses   *26*

1.9   Emerging Threats and Potential Agents of Bioterrorism   *26*
1.9.1   Pandemic Influenza – Human and Avian Influenza Viruses   *26*
1.9.2   Severe Acute Respiratory Syndrome (SARS) – SARS-associated
         Coronavirus (SARS–COV)   *27*
1.9.3   Other Emerging Threats   *27*

Bioterrorism Preparedness. Edited by Nancy Khardori
Copyright © 2006 WILEY-VCH Verlag GmbH & Co. KGaA, Weinheim
ISBN: 3-527-31235-8

2   **Bioterrorism Preparedness:**
          Historical Perspective and an Overview   *33*
          Nancy Khardori
2.1   Introduction   *33*
2.2   International Biodefense Actions in the Nineteenth Century and Their
          Impact   *34*
2.3   Civilian Biodefense – The Obstacles   *36*
2.4   Bioterrorism Preparedness – The Rationale   *39*
2.5   Bioterrorism Preparedness – The Avenues   *40*
2.5.1   Public Health Laws   *40*
2.5.2   Public Health System Preparedness   *42*
2.5.3   Political Preparedness   *52*
2.5.4   Bioterrorism Preparedness – Global Avenues   *53*

3   **Care of Children in the Event of Bioterrorism**   *73*
          Subhash Chaudhary
3.1   Introduction   *73*
3.2   Increased Vulnerability of Children   *73*
3.2.1   Anatomic and Physiological Features Placing Children at Increased Risk of
          Vulnerability   *74*
3.2.2   Developmental Factors Involved in Increased Vulnerability of
          Children   *75*
3.2.3   Delayed Diagnosis in Children   *76*
3.2.4   Unique Management Needs of Children   *76*
3.2.5   Decontamination Showers   *76*
3.2.6   Doses of Medication   *76*
3.2.7   Size of Equipment   *76*
3.2.8   Training of Healthcare Workers to Meet the Special Needs of
          Children   *77*
3.2.9   Communication with Children about Disasters   *77*
3. 2. 10 Communication with Adolescents about Disasters   *77*

3.3   Categories of Biological Agents and Toxins   *78*
3.3.1   Smallpox (Variola)   *78*
3.3.2   Anthrax   *81*
3.3.3   Botulism   *84*
3.3.4   Plague   *87*
3.3.5   Tularemia   *88*

4   **Smallpox: Virology, Clinical Presentation, and Prevention**   *93*
          James M. Goodrich
4.1   Introduction   *93*
4.2   History   *94*
4.3   Virology   *94*

4.4   Clinical Features and Classification   *96*
4.4.1   Rash and Prognosis   *96*

4.5   The Stages of Smallpox   *96*
4.5.1   Incubation Period   *97*
4.5.2   Pre-eruptive Stage   *98*
4.5.3   Eruptive Stage   *100*

4.6   Ordinary Type   *100*
4.6.1   Death   *103*

4.7   Modified-type   *104*
4.7.1   Variola Sine Eruptione and Subclinical Infection   *104*

4.8   Flat-type   *104*
4.9   Hemorrhagic-type   *105*
4.10   Early Hemorrhagic-type   *106*
4.11   Late Hemorrhagic-type   *106*
4.12   Complications   *107*
4.12.1   Skin   *107*
4.12.1   Respiratory   *108*
4.12.1   Gastrointestinal   *108*
4.12.1   Neurological   *108*
4.12.1   Ophthalmic   *108*
4.12.1   Osteo-articular   *108*

4.13   Differential Diagnosis   *109*
4.14   Pathophysiology   *111*
4.15   Laboratory Diagnosis   *112*
4.16   Postexposure Infection Control   *113*
4.17   Vaccination and Immunity   *113*
4.18   Antiviral Treatment   *116*
4.19   Summary   *117*

5   Anthrax – Bacteriology, Clinical Presentations, and Management   *123*
         Nancy Khardori
5.1   Historical Background   *123*
5.2   Epidemiology   *124*
5.3   Microbiology and Genetics   *125*
5.4   Virulence Factors and Pathogenesis   *127*
5.5   Human Anthrax – Clinical Manifestations   *128*
5.5.1   Cutaneous Anthrax   *128*
5.5.2   Gastrointestinal Anthrax   *129*
5.5.3   Inhalational Anthrax   *130*
5.5.4   Hemorrhagic Meningoencephalitis   *132*
5.5.5   Microbiological Diagnosis   *132*
5.5.6   Immunological Tests and Serological Diagnosis   *134*

5.5.7   Antimicrobial Therapy and Post-exposure Prophylaxis   *135*
5.5.8   Emerging/Investigational Therapies   *136*
5.5.9   Human Vaccination   *137*
5.5.10   Anthrax Vaccines in Development   *138*
5.5.11   Infection Control and Decontamination   *139*

6   Plague: Endemic, Epidemic, and Bioterrorism   *147*
         Janak Koirala
6.1   Introduction   *147*
6.2   History   *147*
6.3   Microbiology   *148*
6.4   Global Epidemiology   *148*
6.5   Pathogenesis   *149*
6.6   Clinical Features   *150*
6.6.1   Bubonic Plague   *151*
6.6.2   Primary Septicemic Plague   *151*
6.6.3   Primary Pneumonic Plague   *151*
6.6.4   Other Forms   *152*

6.7   Mortality   *152*
6.8   Laboratory Diagnosis   *152*
6.9   Radiology   *154*
6.10   Potential as a Biological Weapon   *154*
6.11   Features of Bioterrorism   *154*
6.12   Diagnosis   *155*
6.13   Treatment   *156*
6.14   Prevention   *158*
6.14.1   Immunization   *159*
6.14.2   Antibiotic Prophylaxis   *159*
6.14.3   Infection Control   *160*

7   Botulism: Toxicology, Clinical Presentations and Management   *163*
         Janak Koirala
7.1   Introduction   *163*
7.2   History   *163*
7.3   Epidemiology   *164*
7.4   Microbiology and Toxicology   *165*
7.5   Transmission   *168*
7.6   Clinical Features   *168*
7.7   Diagnosis   *169*
7.8   Differential Diagnosis   *171*
7.9   Potential as a Biological Weapon   *172*
7.10   Features of a Botulism Attack   *173*
7.11   Management   *174*
7.12   Prognosis   *174*

7.13   Prevention   *175*
7.13.1 Immunization   *175*
7.13.2 Post-exposure Prophylaxis   *176*
7.13.3 Decontamination   *176*

7.14   Infection Control   *176*

8   Tularemia: Natural Disease or Act of Terrorism   *181*
         Janak Koirala
8.1   Introduction   *181*
8.2   History   *181*
8.3   Microbiology   *182*
8.4   Epidemiology   *183*
8.5   Pathogenesis   *184*
8.6   Clinical Features   *184*
8.7   Laboratory Diagnosis   *186*
8.8   Radiology   *187*
8.9   Potential as Biological Weapon   *187*
8.10   Diagnostic Criteria   *188*
8.11   Treatment   *189*
8.12   Prevention   *191*
8.12.1 Immunization   *191*
8.12.2 Post-exposure Prophylaxis   *191*

8.13   Infection Control   *191*
8.14   Reporting to the Public Health System   *192*

9   Viral Hemorrhagic Fevers:
         Differentiation of Natural Disease from Act of Bioterrorism   *195*
         James M. Goodrich
9.1   Introduction   *195*
9.2   Filoviridae   *199*
9.2.1   Virology   *199*
9.2.2   Epidemiology   *199*
9.2.3   Clinical Manifestations and Disease   *200*
9.2.4   Transmission   *202*
9.2.5   Pathogenesis   *203*
9.2.6   Diagnosis   *204*

9.3   Arenaviridae   *205*
9.3.1   Virology   *205*
9.3.2   Epidemiology   *206*
9.3.3   Clinical Manifestations and Disease   *206*
9.3.4   Transmission   *208*
9.3.5   Diagnosis   *208*

9.4   Bunyaviridae   *208*
9.4.1   Virology   *208*
9.4.2   Epidemiology   *209*
9.4.3   Clinical Manifestations and Disease   *209*
9.4.4   Transmission   *211*
9.4.5   Diagnosis   *211*

9.5   Flaviviridae and Other Viruses   *212*
9.5.1   Kyasanur Forest Disease   *212*
9.5.2   Omsk Hemorrhagic Fever Virus   *212*

9.6   Alphaviruses   *213*

10   Policy Priorities: Smallpox, Stockpiles, and Surveillance   225
         Ross D. Silverman
10.1   Introduction   *225*
10.2   Smallpox Preparedness and Pre-event Vaccination   *225*

**11   Legal Preparedness:
         The Modernization of State, National, and International
         Public Health Law   239**
         Ross D. Silverman
11.1   Legal Preparedness: Sources of Power and Limits   *239*
11.2   Federal Public-health Authority   *241*
11.3   Federal Isolation and Quarantine Powers   *243*
11.4   International Health Regulations   *244*
11.5   Legal Preparedness   *246*
11.6   Legal Preparedness in Action: The Model State Emergency
         Health Powers Act   *247*

Index   *253*

Preface

The range of diseases caused by biological agents and/or their toxins with the potential to be used intentionally against civilian populations is extensive and diverse. Some of these, for example anthrax, have been known to man since antiquity whereas others, for example Nipah virus, were recognized only recently. Even before the "microbial world" was seen or propagated, filth, fomites, carcasses, and cadavers were used to "transmit" disease and devastation to armies during wars.

It is interesting that the first specific biological agent, *Bacillus anthracis*, attributed to human disease by fulfilling Kochs postulates is also the one that has received most notoriety as a bioterrorism agent. The development of the science of bacteriology in the late 19th century expanded the scope of biological agents as weapons of mass destruction. The threat of nuclear and chemical weapons dominated during the 20th century, however. The cheap and easy to propagate biological agents remained in the background and were reported to be used against civilians in isolated incidents mostly by small organized groups or individuals. The United States anthrax attacks of 2001 followed the most devastating and vivid crime against humanity in recent history. The low technology method of successfully disseminating anthrax spores through the US postal service brought into focus the threat of biological agents as potential weapons of mass destruction.

As I looked at the list of diseases caused by "critical biological agents" I immediately realized I had had the opportunity to see a few patients with all of them over the past 32 years. Perhaps this is one of the best things about having had the privilege of working in two different continents and having worked both in the basic science discipline of microbiology and the clinical discipline of infectious diseases. The Infectious Diseases Group (including all the authors of this book) had already planned a regional continuing medical education program in collaboration with the Association of Practitioners in Infection Control (APIC) for November 15, 2001, mostly to address West Nile virus and antibiotic resistance. As the convener, I suggested we expand the scope of the program to include "bioterrorism agents". All parties readily agreed. The program received an overwhelming response and registrations had to be turned down, even after changing the venue to accommodate more delegates. For the first time we were seeing large numbers from all medical and surgical specialties and from specialties like anes-

Bioterrorism Preparedness. Edited by Nancy Khardori
Copyright © 2006 WILEY-VCH Verlag GmbH & Co. KGaA, Weinheim
ISBN: 3-527-31235-8

thesia and radiology in the same room – discussing issues that affected not just their patients but themselves and their families. We were invited by the American Society of Microbiology to conduct the first workshop on bioterrorism at its national meeting on September 26, 2002. We have conducted the workshop every year since in addition to presenting local and regional programs for healthcare providers, hospital executives, and safety engineers.

Last year, I received an invitation from the editor of the second edition of the **Encyclopedia of Molecular Cell Biology and Molecular Medicine** to write a review on *"Preparedness for Bioterrorism"*. As I sent the manuscript, I explained to the editor that the material in this chapter was very different from what I expected to see in other chapters of this encyclopedia. Soon after the materials reached the publishers, Wiley–VCH, I received a very gracious note and an invitation to author and edit a book on bioterrorism. Once again, I chose to depend on my colleagues at our institution and this book is another one of our "team projects".

The book *Bioterrorism Preparedness – A Medicine – Public Health – Policy* has been prepared with the hope of being useful to medical students, healthcare providers, infection control practitioners, public health professionals, and legal professionals involved in health policy issues. The first two chapters provide a historical perspective and overview of potential agents of bioterrorism and bioterrorism preparedness. These two chapters will hopefully provide a quick reference to a variety of issues related to bioterrorism. The third chapter, *"Care of Children in the Event of Bioterrorism"*, has, in my opinion, a unique quality to it. It emphasizes differences between the approach to bioterrorism-related diseases in adults and children – where they exist and are important. The next six chapters (4 to 9) are dedicated to the Category A agents. Each chapter stands on its own and provides appropriate but not overwhelming detail on all aspects of these diseases. The salient features of Category B and Category C agents are discussed in Chapter 1. The last two chapters on policy issues and legal preparedness written by our colleague in the Department of Medical Humanities have truly broadened the scope of this book. It has been a pleasure for me to interact with this young man and recognize the significance of health policy makers in the overall delivery of health care.

As one ponders over the past, present and future of bioterrorism, it becomes clear that the very advances in technology that have made diagnosis and treatment of many infectious diseases possible have also made it simpler to obtain, cultivate, and use them for bioterror. In particular, the breakthroughs that have come from the genomics revolution may be used to enhance detection, protection, and treatment. These same capabilities might also be misused in the design of bioweapons. The threat of biological agents being used for terrorist activity has given an impetus to research that will enhance our capability to detect, trace, and manage bioterrorism events. A significant example of this is the use of genomics in tracing the origin or source of a microbial agent. Microbial forensics will enable "genetic fingerprinting" of the weapon the same way as it is currently being used on the alleged perpetrators. Such research and future technology will at the same time be useful in detecting and managing natural infectious disease. To quote Albert Einstein, "In the middle of difficulty lies opportunity".

I would like to express my sincere thanks to all my colleagues who have made contributions to this book. I must also thank a long time friend and a colleague in endocrinology and molecular medicine who is known for his encyclopedic knowledge, photographic memory, and constant desire to send me reading materials from sources I generally do not follow. In closing, my gratitude and thanks go to Mrs Nancy Mutzbauer without whose unconditional and constant help much of the book would never have seen the light of day.

Springfield, December 2005   *Nancy Khardori*

List of Authors

Subhash Chaudhary
Department of Pediatrics
Southern Illinois University School of
Medicine
P.O. Box 19658
Springfield
Illinois
62794-9636
USA

James M. Goodrich
Pfizer Global Research and
Development
New London
Connecticut
USA

Nancy Khardori
Department of Internal Medicine
Southern Illinois University School of
Medicine
P.O. Box 19636
Springfield
Illinois
62794-9636
USA

Janak Koirala
Division of Infectious Diseases
Department of Medicine
Southern Illinois University School of
Medicine
Springfield
Illinois
62794-9636
USA

Ross D. Silverman
Department of Medical Humanities
Southern Illinois University School of
Medicine
P.O. Box 19603
Springfield
Illinois
62794-9636
USA

Bioterrorism Preparedness. Edited by Nancy Khardori
Copyright © 2006 WILEY-VCH Verlag GmbH & Co. KGaA, Weinheim
ISBN: 3-527-31235-8

1
Potential Agents of Bioterrorism:
Historical Perspective and an Overview

Nancy Khardori

1.1
Historical Perspective – How We Got Here

A quote from Hans Zinser, a bacteriologist and historian during the Great Depression in the United States, puts the concept of "terror associated with biological agents" in the best possible perspective [1]. He said "Infectious disease is one of the great tragedies of living things – the struggle for existence between different forms of life ... incessantly the pitiless war goes on, without quarter or armistice – a nationalism of species against species." What he seemed to convey in this quote is the fact that mankind will never be able to completely protect itself against many of the biological agents coexisting in nature. The interaction between humans and disease-causing pathogens in nature is constant, with one or the other winning at all times and the course of human history has been altered frequently by the capability of infectious agents to spread and cross national borders.

The epidemics and pandemics of infectious diseases caused by communicable agents have swept unchecked across continents claiming more lives and creating more social devastation than wars. Examples include [2]:

1. diseases like smallpox, measles, plague, typhoid, and influenza causing 95 % of deaths in pre-Columbian native American populations;
2. the death of 25 million Europeans (a quarter of the population) caused by plague in the 14th century; and
3. more than 21 million deaths because of the influenza pandemic of 1918 and 1919.

Worldwide, naturally occurring infectious diseases remain the major causes of death. In the United States and Western Europe, the impact of several very virulent microbial agents and/or their toxins has been much reduced because of a very accessible health-care system and the public health infrastructure – although a substantial number of people (approximately 170,000) still die each year from infectious diseases in the United States [3]. The travel and trade necessary for economic globalization, the continued potential for transmission of infectious agents from animals to humans, and large populations living and working in proximity in urban areas of the world enable infectious disease outbreaks to remain a major threat. Recent outbreaks of severe acute respiratory syndrome (SARS) and avian influenza are excellent examples. Until the discovery of preventive measures and anti-infective therapies, for example vaccines and antimicrobial agents, large disease outbreaks were even more common during war times. Infectious diseases caused far more deaths than battle injuries until World War II. Wars led to changes in both the host population of humans and animals and the pathogen population of infectious agents. Humans and animals became more susceptible to disease because of famine and malnutrition and the pathogens found new and vast breeding grounds in decaying organic matter including human and animal corpses. This resulted in pollution of scarce food and water supplies. In addition, vectors, the disease-transmitting agents, for example mosquitoes and flies, multiplied unchecked causing vector-borne diseases for which no preventive measures existed.

It is not surprising that a connection between "disease", "contagion", filth, and foul odor was made much before microbes were discovered. Human ingenuity made use of this association by the crude use of filth, cadavers, and human and animal carcasses as weapons [4]. These avenues of transmitting disease and devastation to armies and civilian populations have been used to contaminate wells, reservoirs, and other water sources since antiquity through the Napoleonic era and into the 20th century. As early as 300 BC, the Greeks polluted the wells and drinking water supplies of their enemies with animal corpses [5]. The same tactics were used later by the Romans and Persians. The bodies of dead soldiers and animals were used to pollute wells during a battle in Italy in 1155. Pollution (poisoning of potable water) was used as an effective and calculated method of gaining advantage in warfare throughout the Classical, Medieval and Renaissance periods. During the Middle Ages military leaders recognized that victims of disease (infections) could themselves become weapons [6]. Gabriel de Mussis, a notary, described how the plague-weakened Tartar forces catapulted victims of plague into the town of Kaffa in 1346 [7]. An epidemic of plague that followed forced a retreat of the Genoese forces. The population under siege may have been at an increased risk of epidemics because of deteriorating sanitation and hygiene. The imported disease continued to spread in Europe. In 1422 bodies of dead soldiers and 2000 cartloads of excrement were hurled into the ranks of the enemy at Carolstein. These two incidents contributed to the 25 million deaths in Europe in the 14th and 15th centuries during the Black Plague. Russian troops battling Swedish forces in Revat resorted to throwing plague victims over the city walls in 1710.

The use of smallpox victims and their fomites as weapons in the new world received similar notoriety. The indigenous people of Central and South America were decimated by measles and smallpox introduced to them by the Spanish conquistadors. They are said to have been presented with smallpox contaminated clothing in the 15th century [6, 8]. Smallpox-laden blankets were provided to the Indians during the French and Indian Wars (1754–1767). This adaptation of the

Trojan Horse use was followed by a smallpox epidemic among native American tribes in the Ohio River Valley. Smallpox epidemics in Native Americans after initial contact with Europeans had, however, been occurring for more than 200 years. Transmission of smallpox by means of respiratory droplets would have been much more efficient than use of fomites. Confederate General Joseph Johnson used the bodies of sheep and pigs in 1863 to pollute drinking water at Vicksburg during the US Civil War. These early attempts (14th to 18th century) at using biological materials to cause disease in the opponent have been referred to as biological warfare even though the nature of the biological agents in these materials was largely unknown. These early incidents also illustrate the complex nature of disease caused by biological agents. Naturally occurring endemic disease is very difficult to differentiate from that caused by deliberate spread of disease. Therefore the concept of "bioterror" should encompass in its spectrum:

1. naturally occurring infectious diseases;
2. acts of biological warfare; and
3. acts of biological terrorism against the civilians in peace and war time.

In any and all of these roles, biological agents have been, and will remain, potential tools of mass casualties.

1.2
Development of Modern Biological Weapons

Bacillus anthracis was the first specific biological agent attributed to human disease when Robert Koch confirmed his own "postulates" concerning this organism in 1877. The subsequent development of the science of bacteriology in the 19th century expanded the scope of biological agents as weapons of mass destruction. This occurred concomitantly with understanding of the pathogenicity of microbes, host–pathogen interactions, and advances in the prevention and treatment of infectious diseases. Modern microbiology intended primarily for diagnosis and treatment of infectious diseases also afforded the capability to isolate and produce stacks of specific pathogens. Germany developed an ambitious biological warfare program during World War I. Covert operations to infect livestock and contaminate animal feed to be exported to the allied forces were conducted in neutral trading partners [9]. *Bacillus anthracis* and *Burkholderia mallei*, causative agents of anthrax and glanders, respectively, were prepared for use to infect Romanian sheep for export to Russia. These cultures were identified at the Bucharest Institute of Bacteriology and Pathology after being confiscated from the German Legation in Romania in 1916. Between 1917 and 1918, livestock in Mesopotamia and Argentina intended for export to Allied Forces were infected with *B. anthracis* and *B. mallei*. During World War I the horror of chemical warfare clearly superceded the impact of biological agents. International diplomatic efforts were directed at limiting the proliferation and use of weapons of mass destruction culminating in the 1925 Geneva Protocol prohibiting the use in war of asphyxiation, poisons, or other gases and of biological methods of warfare [10]. Many of the parties that ratified the Geneva Protocol began research programs to develop biological weapons after World War I. These included Belgium, Canada, France, Great Britain, Italy, the Netherlands, Poland, and the Soviet Union. The United States began an offensive biological program in 1942. Japan conducted twelve large-scale field trials of biological weapons during World War II. This operation was conducted largely under the auspices of Unit 731, a biological warfare research facility. Pathogens used in these experiments included *B. anthracis, Neisseria meningitidis, Shigella* spp. *Vibrio cholera*, and *Yersinia pestis* [11]. During the Japanese program between 1932 and 1945 an estimated 10,000 prisoners died as a result of experimental infection or execution after experimentation. Biological agents were used by Japan to attack 11 Chinese cities. The avenues used included contamination of water supplies and food items, tossing of cultures into homes, and spraying of cultures from aircraft. Pure cultures of *B. anthracis*, V. *cholerae*, Shigella spp., Salmonella spp., and *Y. pestis* were used. Japan was alleged to have used *Y. pestis* as a biological weapon by feeding laboratory bred fleas on plague-infected rats and releasing them over Chinese cities from aircraft. Large numbers of fleas, as many as 15 million, were used per attack to initiate plague epidemics. Rigorous epidemiological and bacteriological data from these experiments are not available. It is estimated that Japan killed 260,000 people in China with biological weapons, primarily plague. Japanese troops suffered approximately 10,000 biological casualties and 1700 deaths, mostly from cholera, in 1941 because they had not been adequately trained or equipped for the hazards of biological weapons. The success of the Japanese attacks attest to the simplicity and diversity with which biological agents can be used to cause death and devastation.

Although the German offensive biological weapons threat during World War II never materialized [12], experiments with *Rickettsia prowazekii, Rickettsia mooseri*, hepatitis A virus and Plasmodia spp. were conducted on Nazi concentration camp prisoners to study pathogenesis and to develop vaccines. As the Weil Felix Test using a cross-reaction immunological method (with Proteus OX19) became available, it was used by the German army to avoid areas with epidemic typhus. As a defense against deportation of people in occupied areas of Poland, physicians used Proteus OX-19 as a vaccine to induce false positivity for typhus. An example of biological weapons being used in a defensive role was created.

The allies developed biological weapon programs for potential retaliatory use in response to German biological attacks. Bomb experiments involving weaponized spores of *B. anthracis* conducted on Gruinard Island near the coast of Scotland, revealed the extensive longevity of viable anthrax spores in the environment. The island was decontaminated with formaldehyde and seawater during 1986 [13]. The United States offensive biological program was begun in 1942 under the direction of a civilian agency, the War Reserve Service [4]. The program weaponized lethal agents such as *B. anthracis*, Botulinum toxin, *Francisella tularensis*, and incapacitating agents such as *Brucella suis, Coxiella burnetii*, Staphylococcus enterotoxin B, and Venezuelan equine encephalitis virus. Anticrop agents such as rice blast, rye stem rust, and wheat stem rust were stockpiled but not weaponized. Cities like New York and San Francisco were surreptitiously used as laboratories to test aerosolization and dispersal methods for simulants. An outbreak of urinary tract infection caused by *Serratia marcescens* occurred at Stanford University Hospital after covert experiments using *S. marcescens* as a stimulant. When the Washington Post reported these covert experiments much later (in 1976) public interest was aroused. The US program was expanded during the Korean War (1950–1953), but the US denied using biological weapons against North Korea and China. The US offensive biological weapons program was terminated after President Nixon's executive orders in 1969 and 1970. Three months later, he extended the ban to include toxins. The US Army Medical Research Institute for Infectious Disease (USAMRIID) at Fort Detrick, Maryland was established to conduct unclassified research on protection against potential agents of bioterrorism.

The origin of the Biological Weapons Program of the former Soviet Union dates back to the statements made by Lenin. Although experimental work was started in the nineteen-twenties, the modern era was ushered in only with the post World War II military building programs [14]. Despite the wide availability of technology for producing and weaponizing biological agents, the direct use of crude fomites against humans continued. One of the examples is the smearing of pungi sticks with excrement by the Vietcong in the early sixties [15]. In 1973 the Soviet Politburo formed the organization known most recently as the Biopreparat to conduct offensive biological weapons programs concealed behind civil biotechnology research [14]. In January 1991 the first ever visit to Biopreparat facilities was undertaken by a joint United Kingdom and United States technical team. By the mid nineteen-nineties substantial changes occurred within the Biopreparat and a concerted effort is in progress to help the Russians civilianize these former biological weapons research and development establishments. The current capability of the old Russian Ministry of Defense sites remains largely unknown. The status of one of Russia's largest and most sophisticated former bioweapons facilities called Vector in Koltsovo, Novosibirsk, is of concern. The facility housed the smallpox virus and work on Ebola, Marburg, and the hemorrhagic fever viruses (e.g. Machupo and Crimean-Congo) [16, 17]. A visit in 1997 found a half-empty facility protected by a handful of guards. No one is clear where the scientists have gone. Confidence is lacking that this is the only storage site for smallpox outside the Centers for Disease Control and Prevention.

Iraq's biological weapons program dates back to at least 1974, started after the Biological and Toxin Weapons Convention had been signed. In 1995, Iraq confirmed that it had produced and deployed bombs, rockets, and aircraft spray tanks containing *Bacillus anthracis* and botulinum toxin [18]. Unfortunately, the number of countries engaged in biological weapons experimentation grew from four in the nineteen-sixties to eleven in the nineties [19]. It is estimated that at least ten nations and possibly seventeen possess biological warfare agents [20]. Of the seven countries listed by the United States Department of State as sponsoring international terrorism, at least five are suspected of having biological warfare programs [21–23]. Nations and dissident groups have the access to skills needed to selectively cultivate some of the most dangerous pathogens and to deploy them as agents of biological terrorism and warfare.

As the technology for cultivating and transporting microorganisms became easier and cheaper, dissident groups and well-financed organizations used biological agents in attacks and threats to accomplish political goals [24, 25]. Some examples of these attempts between 1979 and 2001 are summarized in Table 1.1.

Tab. 1.1
Examples of political attempts at bioterrorism. (Adapted with minor modifications from Ref. [26].)

Year	Group	Attempt	Outcome
1970	Weather Under-ground	A. US revolutionary group intended to obtain agents from Fort Detrick by blackmail and to temporarily incapacitate US cities to demonstrate the "impotence of the federal government"	Report originated with a US Customs informant. The case later seemed to be apocryphal.
1972	R.I.S.E.	A group of college students influenced by ecoterrorist ideology and 1960s drug culture planned to use agents of typhoid fever, diphtheria, dysentery, and meningitis, initially to target the entire world population but later narrowed the plan to five cities near Chicago	The attack was aborted and cultures were discarded
1978	Unknown	Bulgarian defector Georgi Markov was assassinated in London when a spring-loaded device disguised in an umbrella was used to implant a ricin-filled pellet in his thigh.	A similar device used against a second defector in the same area was unsuccessful.
1979	Accidental	Accidental release of anthrax spores from a bioweapons facility in Sverdlovsk, Russia, caused an epidemic of inhalational anthrax.	At least 77 cases and 60 deaths.
1980	Red Army Faction	Members of a Marxist revolutionary ideology group allegedly cultivated botulinum toxin in a safe-house in Paris and planned attacks against at least nine German officials and civilian leaders	This was probably an erroneous report, later repudiated by the German government.
1984	Rajneeshee Cult	An Indian religious cult headed by Rajneesh plotted to contaminate a restaurant salad bar in Dalles, Oregon, with *Salmonella typhimurium*. The motivation was to incapacitate voters, win local elections, and seize political control of the county.	The incident resulted in a large community outbreak of salmonellosis involving 751 patients and at least 45 hospitalizations. The plot was revealed when the cult collapsed and members turned informants.

Year	Group	Attempt	Outcome
1991	Minnesota Patriots Council	A right-wing "Patriot" movement obtained ricin extracted from castor beans by mail order. They planned to deliver ricin through the skin with dimethyl sulfoxide and aloe vera or as dry aerosol against Internal Revenue Service officials, US Deputy Marshals, and local law enforcement officials	The group was infiltrated by Federal Bureau of Investigation informants.
1995	Aum Shinrikyo	A new age doomsday cult seeking to establish a theocratic state in Japan attempted at least ten times to use anthrax spore, botulinum toxin, Q fever agent, and Ebola virus in aerosol form.	Multiple chemical weapon attacks with sarin, Vx, and hydrogen cyanide in Matsumator, Tokyo, and assassination campaigns were conducted. All attempts to use biological weapons failed. The nerve gas sarin killed 12 and injured 5,500 in a Tokyo subway.
1997	Disgruntled employee in Texas	Intentional contamination of muffins and donuts with laboratory cultures of *Shigella dysenteriae*.	Caused gastroenteritis in 45 laboratory workers, four of whom were hospitalized.
2001	Unknown	Intentional dissemination of anthrax spores through the US Postal System led to the deaths of five people, infection of 22 others, and contamination of several government buildings.	Investigation into the attacks so far has not reached a conclusion.

Although most such events do not warrant national or international response and security, they can have substantial public health consequences and therefore require resources and preparedness at the local level. Active surveillance and rapid response at the local level are the cornerstones for preparedness against all types of bioterrorism – "think locally, act globally."

Incidents involving intentional use of microbial agents by small groups or individuals with limited targets are highly likely but the public health consequences are far less. An example is the well publicized arrest on February 18, 1998 of Larry Wayne Harris, a microbiologist who allegedly threatened to release "military grade anthrax" in Las Vegas, Nevada. He had obtained the plague and veterinary vaccine strains of anthrax and reportedly isolated several other bacteria. He made vague threats against US officials on behalf of Christian identity and white supremacists groups. He was arrested when he talked openly about the use of biological agents in terrorist activities. The sensational media coverage appears, however, to have had the unintended effect popularizing anthrax as a potential agent of terrorism among potential perpetrators. The first wave of anthrax hoaxes followed the report of this event. The ease with which he had obtained the cultures prompted new legislation to ensure legitimate medical and scientific purposes for transfer of biological agents.

1.3
Biological Weapons Systems

Acquisition, storage, and transport of biological weapons is much easier than for chemical and nuclear weapons. A biological weapons system comprises:

1. a payload – the biological material consisting of an infectious agent or a toxin produced by bacteria, plants, or animals;
2. munitions that carry and keep the pathogens virulent during delivery;
3. a delivery system, which can be a missile, a vehicle (aircraft, boat, automobile, or truck), an artillery shell, or even an expendable soldier or martyr or conventional mail;
4. a dispersion system that enables dissemination of the payload, in a virulent form, among the susceptible target population [27, 28].

The dispersion system can be aerosol sprays, explosives, and food and water [29]. Aerosol sprays are the most effective means of widespread dissemination and therefore would be the most likely. The factors that can alter the effectiveness of a given dispersion system include the particle size of the agent, stability of the agent under desiccating conditions and ultra violet light, wind speed, wind direction, and atmospheric stability. Optimum conditions and/or the use of hardy organisms would enable clouds of infectious material to travel several hundred kilometers and be delivered to the terminal airways when inhaled. The natural lag time provided by the organism's incubation period (3 to 7 days for most pathogens) would enable safe escape for terrorists before recognition of the attack. Because heat and physical stress inactivate biological activity, explosives are not very effective in disseminating infectious or toxic materials, although the explosion itself and the threat of biological weapons would still create panic, terror, and civil disruption. Effective contamination of large water supplies would usually require an unrealistically large amount of the biological agent. Potable water would be an ineffective dispersion system unless the agent is introduced into smaller reservoirs or into the water supply after it passes through the purification facility. Contamination of food immediately before consumption is easier and more effective in transmitting infectious agents. Unfortunately, an outbreak associated with intentional contamination of food may be recognized late because of difficulty differentiating it from a naturally occurring event. The use of the US Postal service to disseminate anthrax spores carried on pieces of mail has revealed the potential of novel delivery and dispersal systems. Direct delivery of biological agents as pellets and flechettes has also been used. Biological agents can also be used in combination with conventional weapons to create fear and panic, further increasing the potential of mass casualties.

A successful bioterrorism event depends on several factors. For the most optimum outcome:

1. The microbial agents used should have the specific characteristics required of a bioweapon [30, 31]:
    - Most importantly, they must be suitable for mass production, storage, and "weaponization". Transforming microbial agents into bioweapons means they must be able to be packaged and distributed in a manner that disseminates them over a broad area without damaging the pathogenicity, and remain stable during dissemination. Covert release in an urban civilian setting may affect individuals in widely dispersed areas. Although they get the same illness, a common source of infection may not be considered early, because of use of different healthcare providers.
    - They should consistently produce the desired effects of disease and death. These outcomes would be magnified by the fact that both lethal and incapacitating agents would have an adverse impact on civilian health care delivery systems. In a military context, the incapacitation agents may better serve the perpetrator's purpose because the unit will not be able to perform their mission and affected soldiers will consume scarce medical and evacuation resources.
    - They should be highly contagious and infective in low doses. The person-to-person or vector-borne transmission would further increase the number of people affected and enhance the mass casualty effect.
    - They should have a known short and predictable incubation time. This knowledge would favor the terrorists by giving them the lead time and make clinical diagnosis difficult because of multiple possibilities.
    - The disease caused by the agent(s) should be difficult to identify in the target population because of multiplicity of clinical presentation and overlap with common and/or endemic infections. Lack of or low persistence in the environment after delivery would add to the difficulty in determining a "point-source" origin of the disease.
2. The target population should be highly susceptible based on lack of natural or acquired immunity. The lack of herd immunity after infection would lead to ongoing infection as long as the pathogen is around. If no treatment or immunization is available or readily accessible, the disease burden and deaths will increase.
3. The aggressors should have means to protect or treat themselves and their own forces and populations. The presence of partial or full immunity to an agent in the aggressor's population would also be favorable to them.

Biological weapons used in the form of aerosols are invisible, silent, odorless, tasteless and are dispersed relatively easily [32]. They cost 600 to 2000 times less than other weapons of mass destruction. It is estimated that approximately 0.05 % of the cost of a conventional weapon used for biological agents would produce similar numbers of mass casualties per square kilometer [28]. The economic impact of a bioterrorism attack has been estimated to be from $477.7 million per 100 000 persons exposed (brucellosis scenario) to $26.2 billion per 100 000 persons exposed (anthrax scenario) [33].

1.4
Potential Bioterrorism Agents – Categorization and Prioritization

Many potential biological agents are capable of causing human disease. Although bioterrorism attacks could be caused by virtually any pathogenic microorganism, the list of agents that could cause mass casualties by the aerosol route of exposure is very small. Among the diseases caused by agents capable of being "weaponized" are some that are incapacitating while others cause mass casualties. Examples of the latter include anthrax, plague, and smallpox [26]. A North Atlantic Treaty Organization handbook dealing with biological warfare defense lists 39 agents including bacteria, viruses, rickettsia, and toxins as potential agents [34]. The relationship between aerosol infectivity and toxicity versus quantity of agent determine the potential for equivalent effects and narrows the spectrum of possible agents capable of causing mass casualties [23]. For example, only kilogram quantities of anthrax would be needed to cover a 100-km^2 area and cause 50% lethality compared with 8 metric tons of a "highly toxic" toxin such as ricin for similar results. The potential impact on a given area can be determined by the effectiveness of an aerosol in producing downward casualties. In a World Health Organization (WHO) model of a hypothetical dissemination of 50 kg of agent along a 2 km line upwind of a large population center, anthrax and tularemia were shown to cause the highest level of disease and death and the greatest downward spread. Before 1969 both the former Soviet Union and the Untied States spent years determining which pathogens and toxins had strategic and tactical capability. A working group organized by the Johns Hopkins Center for Civilian Biodefense evaluated potential bioterrorism agents to determine which present the greatest risk for a maximum credible event from a public health perspective. A maximum credible event would be one that could cause disruption, panic, and overwhelming of the civilian health-care resources in addition to large loss of life.

Several events in the nineteen-nineties led the US Government to re-embark on a civilian biodefense program [35]. Congress designated Centers for Disease Control and Prevention (CDC) as the lead agency to enhance the nation's epidemiology and laboratory system. A national pharmaceutical stockpile was also established to assist the National Disaster Medical System to manage mass casualties. In addition to its traditional partners (i.e. local and state health departments and laboratories), CDC added the Department of Defense and law-enforcement agencies as its new partners. A Bioterrorism Preparedness and Response Office was established. For initial preparedness five areas were targeted:

1. planning;
2. improved surveillance and epidemiological capabilities;

3. rapid laboratory diagnostics;
4. enhanced communications; and
5. medical therapeutic stockpiling [36, 37].

The biological agents toward which the efforts should be targeted needed to be formally identified and prioritized. A meeting of national experts including academic infectious diseases experts, national public health experts, Department of Health and Human Services representatives, civilian and military intelligence experts and law enforcement officials was convened in June, 1999. Under review were lists of previously identified biological threat agents and potential general criteria for selecting the biological agents that pose the greatest threat to civilians. The lists of potential biological threat agents reviewed included the Select Agent Rule List, Australian Group List for Biological Agents for Export Control, unclassified military list of biological warfare agents, Biological Weapons Convention List and the WHO Biological Weapons List. The general criteria used were:

1. public health impact based on illness and death;
2. delivery potential to large populations based on ability to mass produce and distribute an agent, its stability and potential for person-to-person transmission;
3. public perception of the disease caused by the agent as related to fear and potential civil disruption; and
4. special public health preparedness needs pertaining to stockpiling requirements, diagnostic needs and enhanced surveillance.

Discussions were held to identify agents that were felt to have the potential for high impact based on subjective assessment in these four general categories. After the meeting, CDC personnel tried to identify objective indicators in each category that could be used to further define and prioritize the identified high-impact agents. Rating schemes were used to evaluate agents in each of the general areas according to objective criteria. A risk-matrix analysis process was used to evaluate and categorize potential biological threat agents [37]. The agents were placed in one of three priority categories (A, B, or C) for initial public health preparedness efforts (Table 1.2).

Category A, highest priority agents, include organisms that pose a risk to national security because they:

1. can be easily disseminated or transmitted person-to-person;
2. cause high mortality with potential for major public health impact;
3. might cause public panic and social disruption; and
4. require special action for public health preparedness.

The bacteria, viruses, and toxins listed in CDC Category B are the second highest priority agents; these:

1. are moderately easy to disseminate;
2. cause moderate morbidity and low mortality; and
3. require specific enhancement of CDC's diagnostic capacity and enhanced disease surveillance.

Tab. 1.2
Categorization and prioritization of potential agents of bioterrorism (Adapted with minor modifications from Ref. [26].)

Critical biological agents Category A	Category B	Category C
Variola major (smallpox)	*Coxiella burnetii* (Q fever)	Nipah virus
Bacillus anthracis (anthrax)	*Brucella* spp. (brucellosis)	Hantaviruses
Yersinia pestis (plague)	*Burkholderia mallei* (glanders)	Tickborne hemorrhagic fever viruses
Clostridium botulinum toxin (botulism)	*Burkholderia pseudomallei* (meliododis)	Tickborne encephalitis viruses
Francisella tularensis (tularemia)	Alpha viruses Venezuelan encephalomyelitis Eastern and Western equine encephalomyelitis	Yellow fever
Filoviruses Ebola hemorrhagic fever Marburg hemorrhagic fever	Ricin toxin from *Ricinus communis* (castor beans)	Multidrug-resistant *Mycobacterium tuberculosis*
Arenaviruses Lassa (Lassa fever) Junin (Argentine hemorrhagic fever) Related viruses	*Clostridium perfringens* ε toxin	
	Staphylococcal enterotoxin B T2 mycotoxins	
	Food or waterborne pathogens, including but not limited to: *Salmonella* species *Shigella* species *Escherichia coli* 0157:H7 *Vibrio cholerae* *Cryptosporidium parvum*	
	Rickettsia prowazekii (typhus fever)	
	Chlamydia psittaci (psittacosis)	

Emerging pathogens that could be engineered for mass dissemination are included in the third highest priority Category C. These are:

1. readily available;
2. can be produced and disseminated easily; and
3. have potential for high morbidity and mortality and major health impact.

Ongoing disease surveillance and outbreak response activities are critical to the recognition of diseases caused by emerging pathogens.

In the critical biological agents list, no priority is assigned within the categories. This list does not rank the probability of deliberate use of an agent. Such risk assessments can only be made by intelligence and law enforcement agencies. Although there are severe limitations in predicting the actions of terrorists, risk assessment is critical to balancing preparedness against overreaction.

All Category A critical biological agents will be discussed in Chapters 4–9. A summary is provided in Table 1.3. The overview in this chapter will discuss salient features of the diseases caused by Category B and C agents.

Tab. 1.3
Summary of Category A critical biological agents.

Disease	Agent (type)	Incubation period (mortality without treatment)	Transmission[a]	Disease type	Prevention in humans	Treatment
Smallpox[a]	*Variola major* (virus)	7–10 days (high)	Direct contact Body fluids Fomites	Rash systemic	Vaccine stringent infection control	None (experimental agents)
Anthrax	*Bacillus anthracis* (Bacteria)	1–5 days (high)	Contact – spore Aerosol – spore Contaminated meat	Cutaneous inhalational gastrointestinal meningeal	Vaccine antibiotics	Antibiotics
Plague[a]	*Yersinia pestis* (Bacteria)	1–6 days (high)	Injection – Flea vector Aerosol-droplets	Bubonic pneumonic septicemic	Antibiotics vector protection stringent infection control	Antibiotics
Botulism	*Clostridium botulinum* toxin (Bacterial toxin)	6 h–14 days (high)	Ingestion – food/water Aerosol – toxin	Neurological infantile	None	Antitoxin supportive care
Tularemia	*Francisella tularensis* (Bacteria)	3–14 days (moderate)	Injection – tick vector Aerosol – bacteria Ingestion food/water	Ulcero-glandular systemic	Vaccine vector protection	Antibiotics
Viral Hemorrhagic Fevers[a]	Filoviruses Arenaviruses (viruses)	2–21 days (high)	Body fluids Possible animal reservoir	Hemorrhagic septicemic	Stringent infection control	None

a Person-to-person transmission occurs

1.5
Category B – Bacterial/Rickettsial Agents of Bioterrorism

1.5.1
Brucellosis

Also called undulant fever, Mediterranean fever, or Malta fever.

1.5.1.1 Epidemiology and Microbiology

Brucellosis is a zoonotic disease caused by infection with Brucella, a group of facultative intracellular Gram-negative coccobacilli [27, 38]. Brucella is now regarded as a monospecific genus, that should be termed *B. melitensis*, and other species are subtypes [39]. The natural reservoir is herbivores like goats, sheep, cattle, and pigs. Four subtypes, *B. melitensis* (goat), *B. abortus* (cattle), *B. suis* (pig), and *B. canis* (dog) are pathogenic in humans. Human infection occurs by ingestion of raw infected meat or milk, inhalation of contaminated aerosols, or through skin contact. Brucellosis is highly infective by the aerosol route, with as few as 10–100 bacteria sufficient to cause disease in humans. Brucella sp. are stable to environmental conditions and there is long persistence in wet ground or food. These features favor their use as potential agents of bioterrorism. The disease is relatively prolonged, incapacitating, and disabling in its natural form. Intentional large aerosol doses may shorten the incubation period and increase the clinical attack rate. Mortality rate (5% of untreated cases) is low, however, with rare deaths caused by endocarditis or meningitis. *Brucella suis* became the first agent weaponized by the United States at Pine Bluff Arsenal in 1954, when its biological weapons program was active. Human brucellosis is an uncommon disease in the US. The annual incidence is 0.5 cases per 100,000 population. Most cases occur in abattoir and veterinary workers or are associated with the ingestion of unpasteurized dairy products. The disease is usually seen in Hispanic population and may be related to the illegally imported unpasteurized dairy products from Mexico, where the disease is endemic. The disease is still endemic in many parts of the world (128 cases per 100,000 population in some areas of Kuwait), which makes it a hazard to military personnel in those areas.

1.5.1.2 Diagnosis

The usual incubation period is 8–14 days but may be longer. Brucellosis presents as a nonspecific febrile illness with headache, fatigue, myalgias, chills, sweats, and cough. Lumbar pain and tenderness can occur in up to 60% of cases. Gastrointestinal (GI) symptoms – anorexia, nausea, vomiting, diarrhea and constipation – occur in up to 70% of adult cases, less frequently in children. Hepatosplenomegaly is seen in 45–63% of cases. The significant sequaelae include a variety of osteoarticular infections of the axial skeleton, peripheral neuropathy, meningovascular syndrome, optic neuritis, infective endocarditis, anemia, thrombocytopenia, and leukopenia.

Blood cultures are positive in 15–70% of cases and bone marrow cultures in 92% of cases during the acute febrile phase of the illness. A biphasic culture method (Castaneda bottle) may improve the isolation rate from blood. Because it may take longer to grow Brucella species, the laboratory must be notified to extend the standard incubation time of 5–7 days. If a community laboratory (Level A) observes tiny, faintly staining Gram-negative coccobacilli with slow-growing, oxidase-positive colonies on sheep blood, all plates and bottles should be placed in a biological safety cabinet. They should be appropriately packaged and shipped to a Level B or C laboratory. Confirmation in these laboratories can be done by biochemical, slide agglutination, or phage lysis tests [30].

The diagnosis of brucellosis is frequently made by serological tests. Acute and convalescent phase serum should be collected 3–4 weeks apart. A serum agglutination test (SAT) is available to detect both IgM and IgG antibodies. A titer of 1:160 or greater in a single specimen is considered indicative of active disease. ELISA and PCR methods are becoming more widely available.

1.5.1.3 Management

The United States military recommends doxycycline (100 mg Q12 h) plus rifampin (900 mg a day) for six weeks. Doxycycline for six weeks plus streptomycin for 2–3 weeks is an acceptable alternative. TMP/SMX for 4–6 weeks is less effective. Long-term therapy with a combination of a tetracycline, rifampin and an aminoglycoside is recommended for patients with meningoencephalitis or endocarditis. Valve replacement and surgical intervention for other forms of localized disease may be needed. Chemoprophylaxis is not generally recommended. For a high risk exposure to veterinary vaccine, inadvertent exposure in a laboratory, or exposure as a result of biological warfare, a 3–6 week course of therapy with the agents used for treatment should be considered for prophylaxis.

Live animal vaccines are used widely and have eliminated brucellosis from most domestic animal herds in the US. No licensed human vaccine is available in the US. A variant of *Brucella abortus*, S19-BA has been used in the former USSR. Efficacy is limited and annual revaccination is needed. A similar vaccine is available in China. Because brucellosis is not usually transmissible from person to person, standard precautions are adequate in managing patients. BSL-3 practices should be used for handling suspected Brucella cultures in the laboratory because of the danger of inhalation.

1.5.2
Glanders and Melioidosis

1.5.2.1 Epidemiology and Microbiology

Glanders and melioidosis are caused by *Burkholderia mallei* and *Burkholderia pseudomallei* respectively [38, 40, 41]. Both are Gram-negative bacilli with a "safety pin" appearance on microscopic examination. *Burkholderia mallei*, the causative agent of glanders, produces disease primarily in horses, mules, and donkeys. Human disease is uncommon despite frequent and/or close contact with infected animals. Low concentrations of the organism and less virulence for humans may be the factors responsible. The acute forms of the disease occur in mules and donkeys resulting in death in 3–4 weeks. The chronic form of the disease is more common in horses with lymphadenopathy, multiple skin nodules that ulcerate and drain, along with induration, enlargement and nodularily of regional lymphatics. The later presentation is called "farcy." Human cases occur primarily in veterinarians and animal handlers. Infection is acquired from inhalation or contaminated injuries. *B. pseudomallei*, the causative agent of melioidosis is widely distributed in many tropical and subtropical regions. It is endemic in Southeast Asia and Northern Australia. Humans get infected by inhalation or contact with mucous membranes and skin. Melioidosis is one of the most common causes of community acquired septicemia in Northeastern Thailand. This is a hazard to military personnel in those areas. In humans the disease ranges from a subclinical infection to overwhelming septicemia with 90% mortality rate within 24–48 h. Chronic and life-threatening illness can also occur from reactivation of primary illness.

Aerosols from cultures of *B. mallei* and *B. pseudomallei* are highly infectious to laboratory workers making aerosol spread an efficient way of dissemination. A case of glanders in a military research microbiologist was reported recently [42]. Both of these organisms have been viewed as potential biological warfare agents.

During World War I, glanders was spread deliberately by German agents to infect large numbers of Russian horses and mules. This led to an increase in human cases in Russia after World War I. The Japanese infected horses, civilians, and prisoners of war with *B. mallei* at the Pin Fang (China) Institute during World War II. The United States studied both agents as possible biological weapons in 1943–1944 but did not weaponize them. The former Soviet Union is believed to have experimented with *B. mallei* and *B. pseudomallei* as bioweapons.

1.5.2.2 Diagnosis

The incubation period is 10–14 days. In the acute forms, both glanders and melioidosis can present as an acute pulmonary infection or as an acute fulminant, rapidly fatal septicemic illness. These are the forms that would be expected if they are used as bioweapons. Acute infection of the oral, nasal and conjunctival mucosa can cause bloody nasal discharge with septal and turbinate nodules and ulcerations. Systemic invasion can cause a popular or pustular rash that can be mistaken for smallpox, and hepatic, splenic and pulmonary abscesses. Acute forms of the diseases carry a high mortality rate. The chronic form is characterized by cutaneous and intramuscular abscesses on the legs and arms. Osteomyelitis, meningitis, and brain abscesses have also been reported.

Gram stain of the exudates show Gram-negative bacteria with bipolar staining. They stain irregularly with methylene blue or Wright's stain. The organisms can be cultured and identified with standard bacteriological methods. The cultures should be maintained under BSL-3 conditions. Agglutination and complement fixation tests are available for serological diagnosis of *B. mallei*. Complement fixation tests are more specific and regarded as positive if the titer exceeds 1:20. For *B. pseudomallei*, a single titer above 1:160 with a compatible illness suggests active infection.

1.5.2.3   Management
For localized disease, oral therapy with amoxicillin/clavulanate, tetracycline or TMP/SMX for 60–150 days is recommended. For severe diseases, parenteral therapy with ceftazidime, imipenem or meropenem, plus TMP/SMX for two weeks then oral therapy for six months is recommended. There are no data on post-exposure chemoprophylaxis which may be tried with TMP/SMX, doxycycline, or ciprofloxacin. No vaccine is available for human use. Standard precautions should be used for infection-control purposes. For patients with skin involvement, contact precautions are indicated.

1.5.3
Psittacosis

1.5.3.1   Epidemiology
Psittacosis, caused by *Chlamydophila psittaci*, is a systemic infection with pneumonia as a frequent presentation [43]. It is common in birds and animals. All birds have the potential to spread the disease – ornithosis. However, the name psittacosis has persisted since its association with parrots was described in Greece in 1892. The carriage rate in bird populations is 5% to 8% and bird nesting can shed the organisms during periods of both illness and health. Most human cases of psittacosis, both outbreaks and sporadic cases, occur as a result of contact with a bird (usually a pet). Turkey-associated psittacosis has the highest attack rate in epidemics. Psittacosis is the most common abattoir-associated pneumonia. The infection is transmitted by respiratory route either by aerosolization of infected discharges or by direct contact.

1.5.3.2   Diagnosis
The incubation period for psittacosis is 5–15 days; this is followed by nonspecific symptoms. The illness may resemble a nonspecific viral illness or present as typhoidal and pulmonary syndromes. Atypical pneumonia is the presentation most suggestive of disease caused by *C. psittaci*, which occasionally progresses to acute respiratory distress syndrome. Less common manifestations includes pericarditis, endocarditis, myocarditis, hepatitis, pancreatitis, and thyroiditis. Because the natural route of infection is aerosol, and lungs are the most commonly affected organ from tissue, it would be very difficult to differentiate natural disease from an act of bioterrorism in the early stages.

Diagnosis by culturing the organism from respiratory secretion is possible but hazardous to laboratory workers. Serological diagnosis is made by microimmuno-fluorescence (MIF) by an IgM or IgA titer of 1:16 or by fourfold increase in samples drawn two weeks apart by MIF or complement fixation assay. MIF has higher sensitivity and specificity. Polymerase chain reaction (PCR) with the capability to distinguish *C. psittaci* from other Chlamydia is not routinely available.

1.5.3.3 Management

The antibiotic treatment of choice is tetracycline or doxycyline for 10–21 days. Erythromycin is a less efficacious alternate treatment. Azithromycin, chloramphe-nicol, and newer quinolones have been reported to have activity *in vitro* and in animals. The US Department of Agriculture (USDA) requires that imported birds be quarantined for 30 days and treated with tetracycline for at least 45 days. These requirements are aimed at preventing the introduction of Newcastle Disease and shedding of *C. psittaci* by the birds.

Standard precautions for patient care are recommended, because human-to-human and nosocomial transmission is rare. Antibiotic prophylaxis of contacts in the natural disease setting is not considered necessary. Because the organism is resistant to drying and can remain viable for months at room temperature, environ-mental sanitation is important.

1.5.4
Q Fever

First described in Australia in 1935 and called query fever because the causative agent was unknown.

1.5.4.1 Epidemiology and Microbiology

Q fever is caused by Rickettsia, *Coxiella burnetti* is a world wide zoonosis [27, 38, 44]. The most common reservoirs are cattle, sheep, and goats. Other natural reservoirs are dogs, cats, and birds. The infected animals do not develop the disease but shed large numbers of organisms in body fluids (milk, urine, and feces) and especially large numbers in the placenta. Humans acquire the disease by inhalation of contaminated aerosol. Q fever as a febrile illness with an atypical pneumonia can resemble mycoplasma, legionnaire's disease, *Chlamydia pneumo-nia*, psittacosis, and hantavirus infection. More rapidly progressive cases may resemble tularemia or plague. The organism is resistant to heat and desiccation and highly infectious by the aerosol route. A single inhaled organism is capable of producing clinical illness. *C. burnetti* has the potential to be used as an incapa-citating biological warfare agent and the disease would be similar to that occurring naturally.

1.5.4.2 Diagnosis

The incubation period is 7–21 days, varies according to the number of organisms inhaled. The disease presents as a nonspecific acute febrile illness with headaches, fatigue, and myalgias. Pneumonia, manifested by abnormal chest X-ray occurs in 50% of patients and acute hepatitis develops in 30–60% of patients. Culture negative endocarditis (fewer than 5% of acute cases), chronic hepatitis, aseptic meningitis, encephalitis, and osteomyelitis are uncommon complications of Q fever.

Isolation of the organism is difficult. Coxiella grows in living cells only and cell cultures should be performed under BSL-3 precautions. Antibody assays (IFA and ELISA and complement fixation tests) are available at reference laboratories. IgM antibodies may be detected by ELISA as early as the second week of illness and are diagnostic. The complement fixation test, the most commonly available serological test, is relatively insensitive.

1.5.4.3 Management

All suspected cases of Q fever should be treated to reduce the risk of complications. Tetracycline or doxycycline or erythromycin for 14 days are the treatments of choice for acute Q fever. Azithromycin and clarithromycin would be expected to be effective, although they have not been tested. Ciprofloxacin and other quinolones are active *in vitro* and should be used in patients unable to take the other agents. For endocarditis, tetracycline or doxycycline given in combination with TMP/SMX or rifampin for 12 months or longer has occasionally been successful. Valve replacement is often required for a cure.

Chemoprophylaxis with tetracycline or doxycycline for 5–7 days is effective if started 8–12 days post exposure. If given immediately (1–7 days) after exposure, however, chemoprophylaxis is not effective and may only prolong the onset of disease.

A formalin-inactivated whole cell vaccine is licensed in Australia and available for at-risk personnel on an investigational basis in the US. A single dose provides complete protection against naturally occurring Q fever and greater than 95% protection against aerosol exposure within 3 weeks. Protection lasts for at least 5 years. The vaccine may cause local induration, sterile abscess, and even necrosis at the inoculation site in immune individuals. An intradermal skin test using 0.02 mg vaccine is required to detect presensitized or immune individuals. A live attenuated vaccine (strain M44) has been used in the former USSR. There is no person-to-person transmission of Q fever. Standard precautions are recommended for health-care workers taking care of patients with suspicion or diagnosis of Q fever.

1.5.5
Typhus Fever

Rickettsia prowazekii can cause devastating naturally occurring epidemics of louse-borne typhus [45]. Epidemics are associated with conditions of war, poverty, natural disasters, and lack of hygiene. Typhus has affected the outcome of many wars from the fifteen-hundreds to the 19th century. During World War II, Germany conducted experiments with *R. prowazekii* on Nazi concentration camp prisoners to study pathogenesis and to develop vaccines. *R. prowazekii* is transmitted between patients by the human body louse *(Pediculus humanus corporis)*. This vector is strictly adapted to humans, lives in the clothes and becomes infected while taking a blood meal from Rickettsemic patients. *R. prowazekii* is excreted in the louse feces, deposited on the skin or mucous membrane and introduced by scratching the skin or rubbing the mucous membranes. Latent infection and reactivation of typhus with the potential to start another epidemic can occur. Typhus is endemic in the Peruvian Andes, Burundi, and Rwanda. Recent cases have been reported in Russia, Algeria, Senegal, and France. Southern flying squirrels which are distributed from Florida to Maine and westward to Minnesota and east Texas are an extra reservoir of *R. prowazekii*. Infection can be transmitted to humans by flying squirrel fleas and by exposure to the feces of the fleas or squirrel species of lice.

1.5.5.1 Diagnosis

The incubation period is 8–16 days. After a prodrome of 2 days, rash and fever lasting for 10–12 days occur in about 80% of patients. Severe muscle pains, rigors, malaise, and severe headaches are a part of the clinical picture. Before the availability of antibiotics the course was characterized by hemorrhagic rash, delirium, severe cough, gangrene, coma, and death in 13% of patients. In recent outbreaks in Ethiopia and Burandi the fatality rate was lower because of effective antimicrobial therapy.

Because of overlap of symptoms and signs with many other illnesses, the diagnosis of louse-borne typhus early in an outbreak is challenging. The same would be true for aerosol-transmitted typhus in a bioterrorism event. Epidemiological clues should raise the index of suspicion.

Laboratory methods such as PCR and immunohistochemical detection of *R. prowazekii* in blood and tissue can be used to diagnose the disease during the acute stage. PCR can detect the organism in lice also. *Rickettsiae* can be isolated from blood or tissue in shell vial cell cultures. These diagnostic capabilities are available in reference laboratories only.

Generally, laboratory diagnosis of louse-borne typhus is made retrospectively by serological methods. The methods available include indirect immunofluorescence assay and enzyme immunoassays. An IgM titer of 1:32 and IgG titer of 1:128 confirm the diagnosis. A cross-reacting serological test (Weil–Felix reaction) using *Proteus vulgaris* OX-19 agglutination has poor sensitivity and specificity. This may be the only method available in many parts of the world where typhus is endemic or likely.

1.5.5.2 Management

The antibiotic treatment of choice is doxycycline for 7 to 10 days. Chloramphenicol and tetracycline are also effective. New macrolides, fluoroquinolones, and rifampin have been reported to inhibit the growth of *R. prowazekii* in cell culture. Clinical efficacy has not been proven and treatment failures have been reported with quinolones. The mainstay of prevention of epidemic typhus is the control of body lice. Regular washing of all clothes in hot water stops the outbreak. Insecticides such as lindane powder are useful in delousing. Lice can also be killed by application of 1% permethrin dusting powder to the clothing and bedding every 6 weeks. No vaccine is currently available for prevention of louse-borne typhus.

1.5.6
Food and Water Safety Threats

Food and water borne pathogens as potential agents of bioterrorism include, but are not limited to, the subset in Table 1.4 [46]. "Poisoning" of potable water was used as an effective and calculated method of gaining advantage in warfare throughout the Classical, Medieval, and Renaissance periods. Drinking water supplies of the enemies were polluted with human and animal corpses. During World War I Germany used covert operations to infect livestock and contaminate animal feed to be exported to the allied forces. These operations were conducted via neutral trading partners. The organisms used were *B. anthracis* and *B. mallei*. During biological agent attack on eleven Chinese cities (1932–1945), the Japanese contaminated water supplies and food items. Pure cultures of Salmonella sp., Shigella sp. and *Vibrio cholera* were used. Under current conditions of drinking water treatment and safety, potable water would be an ineffective dispersion system unless the cultures were introduced into smaller reservoirs or into the water supply after it has passed through the purification facility. It is easier and more effective to transmit infectious agents by contaminating food. Such intentional food-borne outbreaks would be difficult to differentiate from naturally occurring events. The spectrum of organisms and of foods causing food-borne disease has expanded in recent years. Many food-borne pathogens, for example salmonella and campylobacter, have become resistant to commonly used antimicrobial agents. In the outbreaks reported to the CDC between 1972 and 2000, the first ten causes of water-borne outbreaks were *Giardia lambia*, Shigella sps., hepatitis A, norovirus, *Campylobacter jejuni*, *Cryptosporidium parvum*, Salmonella sps., *E. coli* 0157-H7, *Salmonella typhi*, and *Vibrio cholerae*.. The ten most common microbial causes of food-borne outbreaks were Salmonella sps, *Staphylococcus aureus*, *Clostridium perfringens*, *Clostridium botulinum*, Shigella sps. *Escherichia coli*, *Campylobacter jejuni*, *Bacillus cereus*, *Vibrio parahemolyticus*, and *Listeria monocytogenes*. There is a significant overlap among water and food-borne pathogens. The evaluation of an outbreak suspected to be food-borne may reveal water to be the vehicle. The diagnosis of food-borne disease should be suspected when two or more persons who have shared a meal during the previous week present with an acute illness with gastrointestinal or neurological manifestations. Important clues to the causative agent are provided by the symptom complexes, the incubation period, the type of food probably responsible, and the setting in which it is consumed. The CDC's

Foodborne Diseases Active Surveillance Network (FoodNet) conducts surveillance for major food-borne pathogens in several states. Prompt reporting by healthcare providers to public health authorities is critical to initiating preventive action including secondary spread of the disease. Because clusters of food-borne disease may arise in geographically different areas in a potential bioterrorism event, intentional contamination of food should be suspected in such instances.

Tab. 1.4
Category B bioterrorism agents.

Bacteria	Viruses	Toxins
1. *Brucella* species (brucellosis)	Alpha viruses – Venezuelan encephalomye-litis	1. Enterotoxin B (*Staphylococcus aureus*)
2. *Burkholderia mallei* (glanders)	– Eastern equine encephalo-myelitis	2. Epsilon toxin (*Clostridium perfringens*)
3. *Burkholderia pseudomallei* (melioidosis)	– Western equine encephalo-myelitis	3. Ricin toxin (*Ricinus communis*)
4. *Chlamydia psittaci* (psittacosis)		4. T2-Mycotoxins[a]
5. *Coxiella burnetti* (Q fever)		
6. *Rickettsia prowazekii* (typhus fever)		
	Food safety threats *Salmonella* species *Shigella* species/dysenteriae *Escherichia coli* 0157:H7	Water safety threats *Vibrio cholerae* *Cryptosporidium parvum*

a Not listed under CDC Category B agents

1.6
Category B – Viral Agents of Bioterrorism

1.6.1
Alphavirus Encephalomyelitis

1.6.1.1 Epidemiology and Diagnosis

Mosquito-borne alpha viruses cause Venezuelan equine encephalitis (VEE), western equine encephalitis (WEE) virus, and eastern equine encephalitis (EEE) [27, 38, 47]. They are similar, share many aspects of epidemiology and transmission, and are often difficult to distinguish clinically. Natural infections are acquired as a result of bites by a wide variety of mosquitoes. In natural epidemics severe and often fatal encephalitis in horses, mules, and donkeys precedes human cases. In a biological warfare attack with the virus disseminated as an aerosol, human disease would be a primary event or occur simultaneously with that in equidae. The human infective dose of VEE is 10–100 organisms. VEE is a febrile, relatively mild incapacitating illness. Encephalitis develops in a small percentage of patients. EEE and WEE viruses cause encephalitis predominately.

1.6.1.2 Management

No specific therapy is available. Alpha-interferon and the interferon inducer poly-ICLC have proven highly effective as post-exposure prophylaxis in experimental animals.

A live attenuated vaccine is available as an investigational new drug. A formalin inactivated vaccine is available for boosting antibody titers in those initially receiving the live attenuated vaccine. The viruses can be destroyed by heat (80°C for 30 min) and standard disinfection. There is no evidence of human-to-human or horse-to-human transmission. Standard precautions and vector control while the patient is febrile are adequate hospital infection control procedures.

1.7
Category B – Biological Toxins for Bioterrorism

1.7.1
Enterotoxin B

Staphylococcus aureus produces several exotoxins, some of which normally exert their effect on the GI tract and are called enterotoxins [27, 38]. These toxins are proteins with a molecular weight of 23,000–29,000 kilodaltons. They are also called pyrogenic toxins because they cause fever. Staphylococcus enterotoxin B (SEB) is a pyrogenic toxin that commonly causes food poisoning originating from improperly handled or improperly refrigerated food. The effect of the inhaled SEB is markedly different. Symptoms occur at a very low inhaled dose (less than one-hundredth of the dose causing GI symptoms). The disease begins rapidly 1–12 h after ingestion with sudden onset of fever, chills, headache, myalgia, and a nonproductive cough. Pulmonary edema occurs in severe cases. GI symptoms can occur concomitantly, because of inadvertent swallowing of the toxin after inhalation. The toxin can also be used to contaminate food or small volume water supplies. SEB was one of the seven biological agents the US bioweapons program possessed before its termination in 1969.

There is no specific therapy available. Experimental immunization has been reported. No human vaccine is available. A candidate vaccine is in advanced development. Secondary aerosols are not a hazard and SEB does not pass through intact skin. Standard precautions for healthcare workers are recommended.

1.7.2
Epsilon (Alpha Toxin)

Clostridium perfringens produces twelve toxins [27]. One or more of these could be weaponized. The alpha toxin, a highly toxic phospholipase can be lethal when delivered as an aerosol. The toxin causes vascular leaks and severe respiratory distress. It can also cause thrombocytopenia and liver damage. The toxin can be detected from serum and tissue samples by a specific immunoassay. Bacteria can be cultured easily. *Clostridium perfringens* is sensitive to penicillin, the current antimicrobial agent of choice. Some data show that clindamycin or rifampin may reduce toxin production by *C. perfringens*. Some toxoids are available for enteritis necroticans in humans. Veterinary toxoids are widely used.

1.7.3
Ricin Toxin

Ricin is a protein cytotoxin derived from the beans of the castor plant (*Ricinus communis*). The castor plant is ubiquitous and the toxin is easy to export. It is stable and highly toxic by several routes of exposure including inhalation [27, 38].

After inhalational exposure, acute onset of fever, chest tightness, cough, dyspnea, nausea, and arthralgia occur within 4–8 h. Acute respiratory distress syndrome in 18–24 h is followed by hypoxemia and death in 36–72 h. Ricin antigen can be detected in the serum and respiratory secretions by ELISA. Retrospective diagnosis is provided by antibody testing in acute and convalescent sera.

No specific therapy is available. Gastric lavage and emetics are recommended after ingestion. Because ricin is a large molecule, charcoal is not useful.

There is no vaccine or prophylactic immunotherapy available for human use. Immunization seems promising in animal models. A protective mask is the best protection against inhalation. Secondary aerosols are not a danger to others and ricin is nonvolatile. Standard precautions are adequate for healthcare workers. Hypochloric solution (0.1 % sodium hypochlorite) inactivates ricin.

1.7.4
T-2 Mycotoxins

Trichothecene mycotoxins are a group of more than forty toxins produced by common molds such as Fusarium, Myrotecium, Trichoderma, Stachybotrys, and other filamentous fungi [27, 38]. They are extremely stable in the environment and the only class of biological toxins that cause skin damage. Hypochlorite solution does not inactivate these toxins. They retain bioactivity even after autoclaving. Skin exposure causes pain, pruritus, redness, vesicles, necrosis, and sloughing. Contact results in severe irritant effects on the respiratory tract, GI tract, and eyes. Severe intoxication results in shock and death. Diagnosis should be suspected if an aerosol attack occurs in the form of "yellow rain" with contamination of the clothes and the environment by pigmented oily fluids.

Treatment is supportive only. Washing with soap and water can prevent or significantly reduce dermal toxicity if done within 1–6 h. Superactivated charcoal should be used after oral intoxication. No prophylactic chemotherapy or immunotherapy is available. Exposure during an attack should be prevented by use of masks and protective clothing. Secondary aerosols are not a hazard. Contact with contaminated skin and contaminated clothing can produce secondary dermal exposures. Until decontamination is accomplished, contact precautions are needed. Subsequently, standard precautions are recommended for healthcare workers. Environmental decontamination requires 1% sodium hypochloride with 0.1 m sodium hydroxide with 1 h contact time.

1.8
Other Toxins With Potential for Bioterrorism

Other toxins include:
- tetanus toxin from *C. tetani*
- toxic-shock syndrome toxin (TSST-1) from *S. aureus*
- exfoliative toxins from *S. aureus*
- saxitoxin – a dinoflagellate toxin responsible for paralytic shellfish poisoning
- tetrodotoxin – a potent neurotoxin produced by fish, salamanders, frogs, octopus, starfish, and mollusks
- toxins from blue–green algae

The agents in Category C, the third highest priority, include emerging pathogens that can be engineered for mass dissemination (Table 1.5).

Tab. 1.5
Category C bioterrorism agents.

Viruses	Bacteria
Nipah virus	Multidrug-resistant
Hanta viruses	*Mycobacterium tuberculosis*
Tickborne hemorrhagic fever viruses	
Tickborne encephalitis viruses	
Yellow fever virus	

1.8.1
Nipah and Hendra Viruses

Nipah and hendra viruses are closely related, were discovered in the last decade, and have been limited to outbreaks in northern Australia, the Malay Peninsula, and more recently Bangladesh [48]. No cases have been documented in the US. Fruit bats (flying foxes) are infected by these viruses but do not show signs of illness. Hendra virus has been reported to cause outbreaks, small clusters, and isolated cases of acute respiratory diseases in humans. There is no evidence of human to human transmission. Horses shed the infectious virus in urine and nasal secretions which may lead to transmission between animals.

Nipha virus caused an outbreak of acute illness in swine and humans between 1998 and 1999 in Malaysia and Singapore. The disease was highly contagious and spread rapidly among swine causing an acute febrile illness, with respiratory symptoms, with a mortality rate of 5 to 15%. The outbreak caused over 1 million deaths in swine. Humans contracted the infection from swine. The illness was more severe in humans and was characterized by fever, headache, myalgia, and encephalitis with 40% mortality rate among the 265 humans. The disease was eradicated from swine in Malaysia with no further outbreaks. The virus is still likely to be present in fruit bats with the potential of reappearing among animals and humans.

Infections caused by hantaviruses and yellow fever virus and the viruses that cause tickborne hemorrhagic fevers are discussed in Chapter 9. Tickborne hemorrhagic fevers include Crimean–Congo hemorrhagic fever, Omsk hemorrhagic fever, and Kyasanur Forest disease [49]. Tickborne complexes of viruses that cause encephalitis in humans include Far Eastern, Central European, Kyasanur forest, Louping ill, Powasan, and probably Negishi [47].

1.9
Emerging Threats and Potential Agents of Bioterrorism

1.9.1
Pandemic Influenza – Human and Avian Influenza Viruses

The threat of global outbreak of influenza is substantial and believed by the WHO and the wider scientific community to be close [50, 51]. The question is whether it will be caused by H5N1, the avian influenza virus strain, or one of the previously known human influenza strains [52]. The recent avian influenza outbreak in east Asia has met two of the three widely recognized prerequisites for a human pandemic:

1. emergence of a new influenza virus (H5N1) against which there is no natural or vaccine-induced immunity; and
2. its transfer to human beings with virulence and remarkably high (72%) mortality.

The virus may even be getting through the final barrier of person-to-person transmission. There is concern that currently circulating H5N1 viruses will evolve into a pandemic strain by adapting to humans by genetic mutation or re-assortment with human influenza strains [53]. These same characteristics favor the use of a microbial agent as a tool for bioterrorism. Avian influenza virus (H5N1) was recently reported to have ocular tropism like the ocular human pathogens adenovirus serotype 37 and enterovirus serotype 70 [54]. Increased surveillance for influenza virus in ocular infections and the use of eye protection when handling avian and zoonotic influenza may reduce bird-to-bird and human-to-human transmission.

1.9.2
Severe Acute Respiratory Syndrome (SARS) – SARS-associated Coronavirus (SARS–COV)

Severe acute respiratory syndrome was first recognized during the outbreak of 2002–2003. It spread from China to more than 50 countries [55, 56]. The pandemic affected over 8000 individuals worldwide and caused over 700 deaths. Soon after identification of this new disease, WHO initiated a global network for collaborative work. A novel coronavirus (SARS–COV) was identified with unprecedented speed and shown to fulfill Koch's postulates. Several epidemiological and public health lessons were learned from the outbreak control measures for SARS in Toronto [57]. The spread of SARS occurred mainly before it was recognized – emphasizing the importance of ongoing surveillance for new and emerging infectious diseases. The importance of good infection-control practices in the healthcare setting, including emergency departments where most cases are seen first, became clear. The onset of a new infection is difficult to detect in the elderly and patients with underlying diseases. The index of suspicion should be higher in the context of an outbreak. Healthcare providers should be aware of the possibility of new disease among returning travelers and their household contacts. The events related to the SARS outbreak emphasize the reality of the times that natural infectious diseases remain a global threat. A global infrastructure for the surveillance of emerging and new pathogens and rapid communications about them is crucial to public health safety.

SARS-COV has several characteristics that make it a concern as a potential tool for bioterrorism. It is highly contagious and lethal; the disease is difficult to differentiate clinically from other respiratory infections, and there is no rapid diagnostic test, treatment or vaccine available. Access to the virus is not heavily restricted as it is for smallpox.

1.9.3
Other Emerging Threats

The recent spread of West Nile virus infection to the US and monkey pox imported via prairie dogs are reminders of what the scientific community has always been concerned about – an infectious disease anywhere in the world is a threat to anyone everywhere in the world. Healthcare workers and public health officials must remain vigilant toward novel or unexplained diseases. The US anthrax attack in 2001 was recognized as a result of meticulous attention to detail by a single clinician. More than 75 % of emerging infectious agents are from zoonotic sources [58]. The history of important environmental factors, e.g. exposure to animals and vectors can be the first clue to a disease outbreak. The magnitude of the zoonotic outbreaks of avian influenza was unprecedented and affected several species of animals [59]. Control of the outbreak has required the implementation of integrated human and veterinary health surveillance and response efforts. These experiences emphasize the value of multidisciplinary approaches to addressing future emerging infectious disease outbreaks, including bioterrorism.

The range of potential weapons of mass destruction and the ways they can be deployed against civilian populations are diverse and extensive [60]. The key elements of protection against biological agents with potential for mass destruction are effective and ongoing surveillance, early detection, and rapid identification. For countermeasures to be effective, they must be deployed before the agent is widely disseminated.

References

1 Zinsser, H. **1963** Rats, lice and history. Boston, MA, *The Atlantic Monthly Press* by Little Brown and Company.

2 Diamond, J. **1999** Guns, germs and steel. New York, NY: www/Norton and Co.

3 Heyman, D. **2000** The global infectious disease threat and its implication for the United States. *National Intelligence Council*; Publication NIEW Washington D. C. 99–170.

4 Christopher, G. W., Cieslak, T. J., Pavlin, J. A. et al. **1997** Biological warfare. A historical perspective. *JAMA*, 278:412–417.

5 Poupard, J. and Miller, L. **1992** History of biological warfare. Catapults to capsomers. *Ann NY Acad Sci.*, 666:9–20.

6 Eitzen, E. M. and Takafuji, E. T. **1997** Historical overview of biological warfare. In: Sidell FR, Takafuju EF, Franz DR, editors. Medical aspects of chemical and biological warfare. *Textbook of Military Medicine Part I, Warfare, Weaponry and the Casualty*. Borden Institute; Washington D. C. 415–423.

7 Derbes, V.J. **1966** De Mussis and the great plague of 1348. A forgotten episode of bacteriological warfare. *JAMA* 96:179–182.

8 Hopkins, D. R. **1983**. Princes and peasants. Smallpox in history. University of Chicago Press, Chicago, Illinois.

9 Hugh-Jones, M. **1992** Wickham Steed and German biological warfare research. *Intelligence and National Security* 7:379–402.

10 Geissler, E. **1986** Biological and toxin weapons: Research, development and use from the Middle Ages to 1945. SIPRI Chemical and Biological Warfare Studies. No. 16. *Oxford University Press Inc*. New York, New York.

11 Harris, S. **1992** Japanese biological warfare research on humans: A case study of microbiology and ethics. *Ann NY Acad Sci*, 666:21–52.

12 Nitscherlich, A., Mielke, F. **1983** Medizin ohne menschlichkeit: Dokuments des Nurnberger Arzteprozesses. *Frankfurt am Main, Germany. Fischer Taxchenbuchverlag.*

13 Manchee, R. J., Steward, R. **1988** The decontamination of Gruinard Island. *Chem. Br* 24:690–691.

14 Davis, C. J. **1999** Nuclear blindness: An overview of the biological weapons program of the former Soviet Union and Iraq. *Emerg Inf Dis*, 5:509–512.

15 Stubbs, M. **1962** Has the West an Achilles heel: Possibilities of biological weapons. *NATO's Fifteen Nations.* June/July, 7:94–99.

16 Henderson, D. A. **1998** Bioterrorism as a public health threat. *Emerging Infectious Diseases,* 4:488–492.

17 Henderson, D. A. **2001** Strengthening global preparedness for defense against infectious diseases threats. *Hearing on the Threat of Bioterrorism and the Spread of Infectious Disease,* 1st Session, 107th Congress.

18 Zalinskas, R. A. **1997** Iraq's biological weapons? The past or future? *JAMA,* 278:418– 424.

19 Rorberts, B. **1993** New challenges and new policy priorities for the 1990's. In: Biologic weapons; weapons of the future. *Center for Strategic and International Studies,* Washington, D. C.

20 Bartlett, J. G. **1999** Update in infectious diseases. *Ann of Inter Med* 131:273–280.

21 Carus, W. S. **1999** Bioterrorism and biocrimes: The illicit use of biological agents in the 20th century. *Center for Counterproliferation Research,* National Defense University, Washington, D. C.

22 Henderson, D. A. **1999** The looming threat of bioterrorism. *Science,* 283:1279–1282.

23 Kortepeter, M. G., Parker, G. W. **1999** Potential biological weapons threats. *Emerg Inf Dis* 5:523–527.

24 Relman, D. A. and Olson, J. E. **2001** Bioterrorism preparedness: what practitioners need to know. *Infect. Med.* 18:497–514.

25 Tucker, J. B. **1999** Historical trends related to bioterrorism: an empirical analysis. *Emerg. Inf. Dis* 5:498–504.

26 Khardori, N. and Kanchanapoom, T. **2005** Overview of biological terrorism: Potential agents and preparedness. *Clinic Microb News* 27:1–8.

27 Stewart. C. **2001** Toxins and biowarfare in biological warfare: Preparing for the unthinkable emergency. Topics in Emergency Medicine. *Am. Health Consult,* Vol II, Atlanta, Georgia

28 Hawley, R. J. and Eitzen, E. M., Jr. **2001** Biological weapons – a primer for microbiologists. *Annu. Rev. Microbiol* 55:235–253.

29 Richards, C. F. et al. **1999** Emergency physicians and biological terrorism. *Ann Emerg. Med.* 34:182–190.

30 Miller, J. M. **2001** Agents of bioterrorism: preparing for bioterrorism at the community health care level. *Infect. Dis. Clin. N. Am.* 15:1127–1156.

31 Beeching, N. J., Dance, D. A., Miller, A. R., et al. **2002** Biological Warfare and bioterrorism. *BMJ* 324:336–339.

32 Danzig, R. and Berkowsky, P. B. **1997** Why should we be concerned about biological weapons. *JAMA* 278:431–432.

33 Kaufmann, A. F., Meltzer, M. I. and Schnid, G. P. **1997** The economic impact of a bioterrorist attack: Are prevention and postattack intervention programs justifiable? *Emerg Inf. Dis* 3:83–94.

34 Departments of Army, Navy and Air Force **1996** NATO Handbook on the Medical Aspects of NBC defensive operations. The Department of Defense., Washington, D. C.

35 Khan, A. S., Moise, S. A., and Lillibridge, S. **2000** Public health prepared-ness for biological terrorism in the USA. *Lancet* 356:1179–1182.

36 Centers for Disease Control and Prevention. **2000**. Biological and chemical terrorism: strategic plan for preparedness and response. Recommenda-tions of the CDC Strategic Planning Workgroup, *MMWR* Vol. 49 No. RR-4.

37 Rotz, L. D., Khan, A. S., Lillibridge, S. R. Ostroff, S. M. and Hughes, J. M. **2002** Report summary. Public Health Assessment of Potential biological terrorism agents. *Emerg Inf Dis* 8:225–229.

38 USAMRIID's Medical Management of biological Casualties Handbook. **2004** U. S. Army Medical Research Institute of Infectious Diseases, Fort Detrick, Frederick, Maryland. 1–117.

39 Pappas, G., Akritidis, N., Bosilkovski, M. And Tsianos, E. **2005** Brucellosis *N Engl J* Med 352:2325–2236.

40 CDC, Division of Bacterial and Mycotic Diseases. **2002** Glanders. http://www.cdc.gov/ncidod/dbmd/diseaseinfo/glanders_g.htm.

41 CDC, Division of Bacterial and Mycotic Diseases. **2002** Melioidosis (*Bu-kholderia pseudomallei*) http?//www.cdc.gov/ncidod/dbmd/diseaseinfo/melioidosis_g.htm

42 Deitchman, S. and Sokas, R. **2001** Glanders in a military research micro-biologist. *N Engl J Med*, 345:1644.

43 Schlossberg, D. **2005**. *Chlamydophila (Chlamydia) psittaci* (Psittacosis) in Mandell, Douglas, and Bennett's Principles and Practice of Infectious Diseases. Churchill, Livingstone, Philadelphia, Pennsylvania. 2256–2258.

44 Dupont, H. T., Raoult, D., Brouqui, P., et. al. **1992** Epidemiologic features and clinical presentation of acute Q Fever in hospitalized patients: 323 French cases. *Am. J Med* 93:427–434.

45 Raoult, D. and Walker, D. H. **2005** *Rickettsia prowazekii* (Epidemic or Louse-Borne typhus) in Mandell, Douglas, and Bennett's Principles and Practice of Infectious Diseases. Churchill, Livingstone, Philadelphia, Pennsylvania 2303–2306.

46 Fry, A. M., Braden, C. R., Griffen, P. M. and Hughes, J. M. **2005** Foodborne disease in Mandell, Douglas, and Bennett's Principles and Practice of Infectious Diseases. Churchill, Livingstone, Philadelphia, Pennsylvania 1286–1301.

47 Whitney, R. J. and Gnann, J. W. **2002** Viral encephalitis: familiar infections and emerging pathogens. *The Lancet* 359:507–513.

48 Ostroff, S. M., McDade, J. E., LeDuc, J. W., Hughes, J. M. **2005** *Emerging and Reemerging Infectious Disease Threats* in Mandell, Douglas, and Ben-nett's Principles and Practice of Infectious Diseases. Churchill, Living-stone, Philadelphia, Pennsylvania 173–192.

49 Borio, L., Inglesby, T., Peters, C. J., et. al. **2002** Hemorrhagic fever viruses as biological weapons. Medical and Public Health management. *JAMA* 287:2391–2411.

50 USAMRIID's Medical Management of biological Casualties Handbook. **2004** U. S. Emerging threats and future biological weapons. Army Medical Research Institute of Infectious Diseases, Fort Detrick, Frederick, Mary-land. 100–108.

51 Oxford, J. S. **2005** Preparing for the first influenza pandemic of the 21st century. *The Lancet* 5:129–131.

52 Osterholm, M. T. **2005** Preparing for the next pandemic. *N Engl J Med* 352:1839–1842.

53 Hien, T., de Jong, M. and Farrar, J. **2005** Avian influenza – A challenge to global health care structures. *N Engl J Med* 351:2363–2365.

54 Olofsson, S., Kumlin, U., Dimock, K. and Amberg, N. **2005** Avian influenza and sialic acid receptors: more than meets the eye? *Lancet Infect Dis* 5:184–188.

55 Editorial **2004** Reflection and Reaction SARS, emerging infections and bioterrorism preparedness. *Lancet Inf Dis* 4:483–484.

56 Groneberg, D. A., Poutanen, S. M., Low, D. E., Lode, H. Welte, T. and Zabel, P. **2005** Treatment and vaccines for severe acute respiratory syndrome. *Lancet Inf Dis* 5:147–155.

57 Svoboda, T., Henry, B., Shulman, L., et al.. **2004** Public Health measures to control the spread of the severe acute respiratory syndrome during the outbreak in Toronto. *N Engl J Med* 350:2352–2361.

58 Molyneux, D. H. **2004** "Neglected" diseases but unrecognized successes – Challenges and opportunities for infectious disease control. *The Lancet* 364:380–383.

59 Witt, C. J. and Malone, J. C. **2005** A veterinarian's experience of the spring 2004 avian influenza outbreak in Laos. *Lancet Inf Dis* 5:143–145.

60 Gosden, C. and Gardener, D. **2005** Weapons of mass destruction – threats and responses. *BMJ* 331:397–400.

2
Bioterrorism Preparedness: Historical Perspective and an Overview

Nancy Khardori

2.1
Introduction

Hans Zinser, a historian and bacteriologist, wrote in 1936, "However secure and well regulated civilized life may become, bacteria, protozoa, viruses, infected fleas, lice, ticks, mosquitoes, and bedbugs will always lurk in the shadows ready to pounce when neglect, poverty, famine, or war let down the defenses. And even in normal times they prey on the weak, the very young and the very old, living along with us, in mysterious obscurity waiting for opportunities." Almost seven decades later Madeline Drexler wrote, "Modern adventurers like to up the ante, but even the most extreme sports wouldn't produce the adrenaline of a race against pandemic influenza or a cloud of anthrax at the Super Bowl. In the field of infectious diseases, reality is stranger than anything a writer could dream up. The most menacing bioterrorist is Mother Nature herself" [1]. Although there have been incidents of intentional use of infectious agents against civilians in the recent years, the "natural" outbreaks of already known human diseases such as Ebola and Marburg virus hemorrhagic fevers and previously unrecognized diseases like severe acute respiratory syndrome (SARS) and avian influenza have had significantly larger impacts on public health and public health resources. In fact, more than 30 new infectious diseases have been described in the last 30 years [2]. Worldwide, influenza experts recognize the inevitability of another pandemic [3]. The influenza pandemic of 1918 and 1919 was shown by a recent analysis to have killed 50 to 100 million people. Today, the world population is more than three times that in 1918. Therefore, even a relatively "mild" pandemic could result in millions of deaths. On the basis of current figures for the reported cases, the mortality rate from avian influenza is remarkably high – more than 72 % compared with an estimated 2.5 % for Spanish influenza [4]. Many modern demographic and ecological conditions favor the spread of infectious disease within and across continents [5]. The most significant factor is the ease and frequency of movement across international boundaries for business, tourism, immigration, etc. Other important factors include rapid population growth, increasing poverty, urban migration, alterations in the habitats of animals and arthropods that transmit disease to humans, aging world population, increasing number of people with impaired host defenses, and changes in food processing and distribution. Emergence of drug-resistant forms of common pathogens and the potential of genetically engineered organisms with unpredictable sensitivity to antimicrobial agents makes global control of infectious diseases even more difficult and complex. Advances in preventive, diagnostic, and pharmacological intervention are needed to protect the population from emerging and re-emerging infectious disease including those related to potential bioterrorism events. The public health system must be prepared for the unexpected and adopt the principle to "Think locally, act globally." The partnership of health-care providers with the public health system forms the foundation of preparedness against natural infectious diseases and the intentional use of biological agents. Cooperation between public health, veterinary medicine, entomologists, and epidemiologists is crucial to monitoring disease prevalence and its impact on human life. Effective understanding and communication between the health care systems and law enforcement and security agencies ensures implementation of policies and reduces the potential dissemination of bioterror – real or perceived.

2.2
International Biodefense Actions in the Nineteenth Century and Their Impact

1. Recorded events pursuing defense actions against non-conventional warfare started in 1899 with the Hague Convention [6]. Approximately two dozen countries signed a pledge not to use toxic gases or other poisons as weapons. This was followed by the use of chemical and biological weapons in combat in 1910–1920 [7].

2. The use of biological and chemical weapons in war was prohibited by the Geneva Protocol in 1925. The United States signed but did not ratify the treaty. This was followed by robust biological weapons programs by Germany and alleged use of *B. anthracis* and other lethal and incapacitating agents during World War II.

3. Post World War II military building programs were started by the former Soviet Union and the United States. The allied biological weapons program shifted from the British World War II anthrax cattle cake retaliation weapon to a large US-based research, development, and production capability.

4. In July 1969 Great Britain submitted a proposal to the United Nations Committee on Disarmament prohibiting the development, production, and stockpiling of biological weapons. The proposal provided for inspections in response to alleged violations [8]. The Warsaw Pact nations submitted a biological disarmament proposal without provisions for inspection

in September 1969. In November 1969 President Richard M. Nixon unilaterally renounced the use of biological weapons in war by the United States. Three months later the ban was extended to toxins. Subsequently, the World Health Organization issued a report on the potential consequences of biological warfare and made estimates of casualties from hypothetical biological attacks. It was estimated that release of 50 kg of agent by aircraft along a 2 km line upwind of a population center of 500,000 would cause 95,000 deaths by anthrax, 30,000 deaths by tularemia, 19,000 deaths by typhus, and 9,500 deaths by tick-borne encephalitis.

5. The Biological and Toxin Weapons Convention (BWC) was developed in 1972. The BWC prohibits the development, possession, and stockpiling of pathogens or toxins in "quantities that have no justification for prophylactic, protective or other peaceful purposes." The treaty also prohibits the development of delivery systems and requires parties to destroy stocks of biological agents, delivery systems, and equipment within 9 months of ratifying the treaty. The BWC treaty was opened for signature in Washington, London, and Moscow on April 10, 1972. There were more than 100 signatory nations including the United States, the Soviet Union, and Iraq. The US Congress ratified the Biological and Toxin Weapons Convention and the 1925 Geneva protocol on January 22, 1975. The BWC entered into effect on March 26, 1975. Review conferences for BWC were held in 1986, 1991, and 1996. There are now 143 state parties and an additional 18 signatories to the Convention [9]. Unfortunately, the BWC was not accompanied by effective provisions for verification. Article IV of the Convention has proved to be an inadequate mechanism for actions against noncompliance.

6. The Soviet Politburo formed the organization known most recently as Biopreparat in 1973. The organization was established to conduct offensive biological weapons programs concealed behind civil biotechnology research. The former Soviet Union had signed the BWC at its inception in 1972. Iraq's biological weapons program also started in 1974 after the BWC had been signed.

7. Initiatives in the 1980's failed to curb biological and chemical weapons proliferation. Iraq used chemical weapons against Iran during the eight year war and also against dissident Kurdish communities in Iraq.

8. Efforts to improve compliance with the BWC began in 1991. The Russian President Boris N. Yeltsin declared the discontinuation of Russia's biological weapons program in April 1992. Iraq confirmed that it had produced and deployed bombs, rockets, and aircraft spray tanks containing *B. anthracis* and botulinum toxin in 1995.

9. In the US new regulations to limit access to chemicals and pathogens that could be developed into weapons (the Select agents) came into effect under the Antiterrorism and Effective Death Penalty Act of 1996.

10. The "dual-use" of infectious agents for legitimate scientific purposes and as potential bioweapons, including those made in biodefense facilities, has made it challenging to police compliance. As information about the covert

biological weapons programs of the former Soviet Union and Iraq became public, provisions for improving verification in the BWC gained momentum [10]. An international ad hoc group of over 50 countries including the USA worked for over 10 years on the BWC verification protocol [11]. Under the protocol the convention signatories would declare the "dual-use" facilities, which would then be subject to random "transparency visits" or "clarification visits" (should questions arise) or "investigation" (if noncompliance is suspected) [10, 11]. In July 2001 the US rejected the Protocol and the whole approach to the verification process. This was a major setback to the 2001 BWC review conference and to the international biological weapons program. The conference was prematurely suspended and reconvened on November 11, 2002, for four days only. No significant objectives were accomplished and a series of annual meetings with narrow agendas were scheduled. The next Convention Review Conference will be held in 2006. As in the Chemical Weapons Convention (CWC), a strong bioweapons verification protocol within the framework of BWC would add to the deterrence of bioweapons which are a much greater threat than chemical weapons.

2.3
Civilian Biodefense – The Obstacles

The most important factors preventing a "safer world" and which are root causes of any type of terrorism include poverty and inequality, enduring state failure, war, human rights abuses, dispossession, and environmental degradation [11]. Environmental degradation coupled with lack of or failing public health infrastructure in most parts of the world contribute significantly to the fact that infectious diseases remain the major causes of morbidity and mortality.

A community registry of births and deaths and a community-based early warning surveillance and response system for priority infectious diseases were established by WHO in Uganda during the Ebola outbreak [12]. This simple and cost-effective measure gave the area the capability to detect new cases of hemorrhagic fever by local staff. Unfortunately most parts of the world still lack the capacity for epidemic detection and response. The lack of local vigilance and preparedness in any part of the world is a threat to global health security. If the capacity of countries to respond even to natural infections is weak, it obviously would be much more difficult to respond to and manage potential bioterrorism events [13].

In the US, the National Association of Counties conducted a survey of county public health directors about their level of preparedness to respond to a bioterrorism or chemical warfare event [14]. Among the responders (31%) the sample size represented 300 counties in 36 cities. Less than 10% of all responding counties said that they were fully prepared. Approximately 70–80% reported being partially prepared and more than 20% did not consider themselves being prepared. The highest level of no preparedness (56%) was reported by counties with population less than 10,000. Among the obstacles cited were insufficient funding (42%), insufficient medical staff (40%), insufficient administrative staff (45%), and insufficient communication networks (35%). One half of the responding counties had no or insufficient policies or procedures in place to enforce quarantine in the event of a bioterrorism event. Results of a survey before September 11, 2001, suggested there was little collaboration between West Virginia County Health Department and local hospitals in preparing to respond to a weapon of mass destruction [15]. More than 60% indicated that primary responsibility for identifying biological agents rested in an agency other then the county health department.

DiGiovanni et al. studied the community reaction to a simulated intentional aerosolized release of Rift Valley fever virus in a semi-rural community in the Southern part of the US [16]. Journalists were considered key participants in the study because of their role in communicating risks. All groups involved in the exercise put more trust in and demanded information from local sources. Members of the media were, however, more fearful than any other group of first responders, other than spouses, made high demands for vaccines, had the poorest understanding of the medical issues, and were most likely to stay away from work after terrorism was recognized. The reaction of members of the community suggested that bioterrorism training should include information management communicators and public affairs officers.

The Council of State and Territorial Epidemiologists (CSTE) compared epidemiologic capability, including terrorism preparedness and response in state health departments, between November 2001 and May 2004 [17]. Federal funding for state public health preparedness programs increased from $67 million in fiscal Year 2001 to approximately two billion dollars during 2002–2003. Although survey results indicated an overall improvement in terrorism epidemiologic and surveillance capacity, to meet federal terrorism preparedness program requirements a need for further increase in epidemiologists and other resources was expressed. The dual use of terrorism and emergency preparedness epidemiology resources was recommended by CSTE to prepare for and respond to terrorism, infectious disease outbreaks, and other public health threats. Few countries in the world have the capacity to spend amounts close to what the US is spending. The lack of "biological literacy and awareness" among policy makers is also a serious obstacle, particularly appreciation of just how different biological attacks are from traditional threats.

Several bioterrorism exercises have been conducted in the US in recent years [18–22]. These include:

1. Operation TOPOFF conducted in 2000 and named for its engagement of the top officials of the US government. Local, state, and federal officials and the staff of three hospitals in metropolitan Denver participated in the exercise using *Y. pestis*. The problematic issues of leadership and decision making, the difficulties of prioritization and distribution of scarce resources, and crisis in the health care settings dealing with triage and spread of disease became very clear early on during the exercise.

2. Shining light on "Dark Winter". This exercise was conducted in the summer of 2001 by the Johns Hopkins Center for Civilian Biodefense strategies in collaboration with the Center for Strategic and International Studies, the Analytic Services Institute for Homeland Security, and the Oklahoma National Memorial Institute for the Prevention of Terrorism. This was a senior level "table top" exercise that simulated a covert smallpox attack on the United States. The exercise offered instructive insights and lessons for those with responsibility for bioterrorism preparedness in the medical, public health, policy, and national security communities. Unfortunately many of the challenges and difficulties faced by the Dark Winter participants were paralleled in the response to the anthrax attacks of 2001.

3. The Hanuman Redux 2001 exercise was conducted by the CDC in Louisville, Kentucky, using *Francisella tularensis* as the biological agent. The conclusion of the exercise was that regions with adequate bioterrorism response plans can respond to a sizeable attack and that local, state and federal health, law enforcement, and other agencies can work together effectively.

4. TOPOFF 2, a multimillion dollar federally funded bioterrorism exercise targeting Chicago and Seattle was conducted in May 2003. The drill was headed by the Department of Homeland Security, and involved more than 8500 people from 100 federal, state and local agencies, the Red Cross, and the Canadian government. The total cost was approximately $16 million. The exercise enabled testing of several preparedness plans put in place after September 2001. The full-scale TOPOFF 3 exercise, managed by the US Department of Homeland Security, took place from April 4–8, 2005. Numerous federal departments and agencies, the states of Connecticut and New Jersey, the United Kingdom, Canada and representatives from the private sector were involved in the exercise.

The exercise, Atlantic Storm, was organized by the Center for Biosecurity, University of Pittsburg Medical Center and the Center for Transatlantic Relations, Johns Hopkins University [23]. A simulated smallpox attack on several cities in Europe and the US was studied. The results showed that Europe, Canada and the US are not sufficiently prepared to counter a major bioterrorism attack. On the basis of results from Atlantic Storm no international organization, including NATO, the United Nations, or the European Union could be relied upon to handle the challenges posed by an attack that penetrates several nations.

Recent advances in biotechnology have enabled widespread use of genetically altered microorganisms to create vaccines, purify proteins, construct cloning vectors, study pathogens, and understand complex host–pathogen interactions [24]. The same technology can be used to enhance antibiotic resistance of biological agents, modify their antigenic properties and transfer virulence factors between them [25]. The techniques of molecular biology can also be used to genetically engineer disease vectors. This "dual use" capability of biotechnology is feared to potentially create "enhanced" or customized biological agents that could cause infectious disease outbreaks and/or be used as bioweapons [26]. Such biological agents would have characteristics that increase their threat as infectious agents, because of enhanced resistance to antimicrobial agents, increased pathogenicity, ability to evade natural or vaccine-induced immunity, failure to be detected by available tools, and altered transmission properties [24]. The first genetically altered microbe for use as a weapon was reportedly developed in 1983 at the State Research Center for Applied Microbiology in Obolensk, a premier former Soviet Union Biopreparat Facility [27]. The alleged bioweapon, reported by Dr Vladimir Pasechnik who defected from the former Soviet Union in 1989, is a hypervirulent strain of *Francisella tularensis*, the causative agent of tularemia. Dr Ken Alibek, another defected Soviet scientist, has described research efforts by Soviet scientists to insert the genes of Venezuelan equine encephalitis virus, β-endorphin, and Ebola virus into smallpox virus and to develop *Bacillus anthracis* (the causative agent of anthrax) strains that can resist vaccine-induced immunity and commonly used antibiotics [28]. Many other examples of recombinant pathogens created at a variety of facilities in the former Soviet Union have been described by other scientists [29]. The actual disease-causing potential of these recombinant or altered pathogens is unknown. Security experts are concerned that the genetically altered strains remain in Russia under uncertain biosafety and biosecurity conditions. This may lead to accidental release and disease outbreaks similar to the anthrax outbreak in Sverdlovsk (now Ekaterinburg, Russia) in 1979. Such dangerous agents created by research can be acquired by terrorist groups or diverted for non-scientific purposes [30]. Even unintentional release of such microbes into the environment has the potential to cause "natural outbreaks" in the future.

2.4
Bioterrorism Preparedness – The Rationale

Bioterrorism has been defined by the CDC as "the intentional release of viruses, bacteria, or toxins for the purpose of harming or killing civilians" [31]. Some states (e.g. Colorado) had legislative and others (e.g. Rhode Island) had administrative public health response plans for a bioterrorism event before September 11, 2001 [32]. The law is a vital component of the public health infrastructure and has been regarded an important tool of public health. In the aftermath of events of September–October 2001, the CDC requested the Center for Law and the Public's Health at Georgetown and Johns Hopkins Universities to draft the Model State Emergency Health Powers Act (MSEHPA or the Model Act) as a part of the process to strengthen the public health infrastructure in the US. The Model Act was drafted in collaboration with members of national organizations representing governors, legislators, attorneys general, and health commissioners. The authors of the act define bioterrorism as "the intentional use of a pathogen or biological product to cause harm to human, animal, plant, or other living organism to influence the conduct of government or to intimidate or coerce a civilian population" [32]. The

National Intelligence Council for the Central Intelligence Agency reported that infectious disease is not only a public health issue but also an issue of national security.

Consequent to the rapid mass casualty effect, biological weapons can overwhelm services and the health care system of communities. Most of the civilian population in the US is susceptible to the infections caused by these agents, resulting in high morbidity and mortality. The economic impact of a biological attack has been estimated to be \$26.2 billion/100,000 persons with potential exposure to anthrax. The economic impact would be greatly enhanced by the resources needed to decontaminate the environment, depending on the pathogen. The estimated cost of decontaminating parts of the Hart Building alone in Washington D. C. after the 2001 anthrax attacks through the postal service was reported to be \$23 million. The model described by Kauffman et al. [33] provides justification for prevention and post attack intervention programs. Improving the public health infrastructure and preparedness against potential agents of bioterrorism would significantly strengthen global defense against infectious disease threats in general.

2.5
Bioterrorism Preparedness – The Avenues

2.5.1
Public Health Laws

The need for state, federal and international public health law reform has come into clear focus because of bioterrorism concerns and naturally emerging infectious diseases, both with the potential of mass casualties. In the US the power to act to preserve the public's health is constitutionally reserved primarily to the states as an exercise of their "Police Powers." In the aftermath of September 11, 2001, The Model State Emergency Health Powers Act (MSEHPA) or the Model Act was drafted and designed to update and modernize the State Public Health Statutes and to avoid problems of inconsistency, inadequacy and obsolescence [32]. The act is structured to facilitate five basic public-health functions:

1. preparedness, comprehensive planning for a public health emergency;
2. surveillance, measures to detect and track public health emergencies;
3. management of property, ensuring adequate availability of vaccines, pharmaceuticals and hospitals and providing power to abate hazards to the public's health;
4. protection of persons, powers to compel vaccination, testing, treatment, isolation, and quarantine when clearly necessary; and
5. communication, providing clear and authoritative information to the public.

The MSEHPA creates the conditions for public health preparedness (e.g. planning, surveillance, and communication). Concerns have been raised about the impact of the Model Act on civil liberties and personal rights [34]. Compulsory power has always been a part of Public Health Law, the MSEHPA actually provides checks and balances against government abuses, clear standards for the exercise of power, and rigorous procedural due process. As of July, 2004, 33 states and the District of Columbia had passed bills or resolutions containing provisions from the Model Act. It will still be necessary for the states to develop contingency plans and to conduct training for judiciary, public health, and health-care providers. Some authorities argue that federal rather than state authorities should design and manage response to bioterrorism. This will ensure larger financial resources and avoid confusion under circumstances demanding emergent response and intervention.

In 1996, the US Defense Against Weapons of Mass Destruction Act designated the Department of Defense as the lead agency to enhance domestic preparedness for responding to and managing the consequences of terrorists' use of weapons of mass destruction [19]. The "Select Agent Program" regulations became effective on April 15, 1997. This act mandated registration with the CDC of laboratories that transfer or receive select biological agents (listed in Section 72.6 of Title 42, Code of Federal Regulations) [35]. The US Patriot act was passed on October 25, 2001 and signed into law the following day. This act amends the Biological Weapons Statute and criminalizes possession of such materials of a type or in a quantity not reasonably justified by bona fide research or peaceful purpose. It also prohibits possession by "restricted persons" in a number of categories set forth in the Act. On July 21, 2005, the US Congress passed a bill that would make permanent several controversial provisions of the Patriot Act. The Public Health Security and Bioterrorism Preparedness and Response Act of 2002 was passed by the US Congress on May 23, 2002 and signed into law June 12, 2002. The Act was designed to improve coordination of federal antibioterrorism activities, improve the health-care system's ability to respond to bioterrorism, protect the nation's food supply and drinking water, address shortages of specific types of health professionals, and speed up the development and production of new drugs and vaccines. The act also increases investment in federal, state, and local preparedness and expands controls over the most dangerous biological agents and toxins. The American Society for Microbiology (ASM) worked closely with congress in drafting Title II of this law, to balance public health concerns over safety and security with the need to protect the use of biological agents in legitimate scientific research and diagnostic testing. New provisions for the possession, use, and transfer of Select Agents (42 biological agents and toxins in listed in Appendix A of 42 CFR Part 72) led to enhanced controls on dangerous biological agents and toxins [36]. The law requires all persons possessing biological agents or toxins deemed a threat to public health to notify the Secretary, Department of Health and Human Services (HHS). The law also requires all persons possessing biological agents or toxins deemed a threat to animal or plant health and to animal or plant products to notify the Secretary, United States Department of Agriculture (USDA). Both secretaries are to be notified when a person possesses agents that appear on both the HHS and the USDA list of Agents and Toxins – designated as HHS/USDA overlap agents (Appendix A). The CDC and the Animal and Plant Health Inspection Service

(APHIS), respectively, were designated as the HHS and USDA agencies responsible for implementing the law [36]. Addressing the challenges posed by a catastrophic health event will need law and order but cannot be solved by them. It will require a robust public health infrastructure to conduct essential public health services at a level that matches the constantly evolving natural and intentional biological agent threats to the health of the public.

2.5.2
Public Health System Preparedness

2.5.2.1 Emergency Response Capability

In the US the National Disaster Medical System has voluntary access to approximately 100,000 hospital beds across the country to cope with a large-scale medical emergency [37]. Not all have the facilities and equipment for mechanical ventilation and other specialized supportive care that may be needed for seriously ill people, however. Such equipment is not available in large quantities even from the Department of Defense War Stacks [38]. Because current federal plans favor freeing up local bed space for injured/affected people, localities need to increase their own capabilities [39]. The Office of State and Local Domestic Preparedness support within the Department of Justice developed a set of objective criteria to measure domestic readiness for a mass casualty event. No locality was found to be prepared for such a crisis [40]. The CDC has established a cooperative grant program with states and several large cities that focuses on bioterrorism preparedness activities [41].

2.5.2.2 Recognition and Training

Rapid response to rare conditions and evaluation of unusual clusters of disease hinges on effective surveillance systems. This is what led to the creation of the US Epidemic Intelligence Service (EIS) which has provided training and disease detection expertise for more than 50 years. The recent influx of emergency response funds has substantially strengthened the epidemiological capabilities of the public health system in the US [42]. Despite this increased funding, state and territorial health departments report that a 47% increase in the number of epidemiologists is needed to fully perform the nation's essential public health services that depend on epidemiology. A large proportion of the current state and territorial epidemiology work force lacks formal training in epidemiology.

Because potential attacks with biological agents are more likely to be covert and unannounced, they pose different and untested challenges to the emergency response system. The traditional "first responders" are prepared for overt attacks like bombings and chemical agents that cause immediate and obvious effects. Because of the incubation period involved, the attack by a biological agent will most probably not have an immediate impact. Therefore, the first victims of a bioterrorism event will need to be identified by health-care workers, who must be aware of the symptoms and epidemiologic patterns of diseases that are otherwise not seen in the US population. In 1997, the Department of Defense received 36 million dollars that were used to initiate the Domestic Preparedness Program to enhance existing first-responder training in dealing with terrorist incidents involving radiological, nuclear, chemical, and biological weapons. The program trained fire departments, law enforcement, hazardous materials, and emergency medical personnel in the 120 largest cities in the US. The personnel were given six separate courses during a week of training [43, 44]. The program was later turned over to the Department of Justice.

When the first wave of victims has been identified, public health officials will need to determine that an attack has occurred, identify the likely organism and prevent more casualties through prevention strategies, e.g. mass vaccination, prophylactic treatment, and infection control procedures. The clues to a potential bioterrorism attack are listed in Table 2.1.

Tab. 2.1
Clues to a potential bioterrorism attack.

Epidemiological clues

Greater than expected case load of a specific disease.
Unusual clustering of disease for the geographic area.
Disease occurrence outside normal transmission season.
Simultaneous outbreaks of different infectious diseases.
Disease outbreak in humans after recognition of disease in animals.
Unexplained number of dead animals or birds.
A disease requiring for transmission a vector previously not seen in the area.
Rapid emergence of genetically identical pathogen from different geographical areas.

Medical clues

Unusual route of infection.
Unusual age distribution or clinical presentation of common disease.
More severe disease and higher fatality rate than expected.
Unusual variants of organisms.
Unusual antimicrobial susceptibility patterns.
A single case of an uncommon disease.

Miscellaneous clues

An intelligence report.
Claims of a release.
Discovery of munitions or tampering.
Increased numbers of pharmacy orders for antibiotics and symptoms relief drugs.
Increased number of 911 calls.
Increased number of visits with similar symptoms to emergency departments and ambulatory healthcare facilities.

It is now recognized that emergency departments and urgent care centers are in the frontline for early recognition of emerging public health threats including those posed by bioterrorism agents. Initial detection, response, and management will always be local. Therefore, periodic education of health care workers, preparedness exercises by the local health departments, and disaster drills by healthcare institutions have become a part of bioterrorism preparedness programs. Primary care physicians played an important role in responding to the 2001 anthrax attacks [45].

2.5.2.3 Outbreak Investigations, Surveillance and Disease Reporting

Outbreaks with unexpected or unusual epidemiological or clinical characteristics should be investigated bearing in mind the possibility of intentional use of biological agents [46]. The CDC is often notified about outbreak investigation by a state or national health department. The origin of such reports is diverse and ranges from private citizens to healthcare providers, local and state health departments, the US Food and Drug Administration, and the World Health Organization. Reports recorded as originating from local or state health departments are usually brought to the attention of the health departments by frontline practitioners. Because of the crucial importance of early detection and reporting, bioterrorism preparedness efforts must include education and support of health care professionals including clinicians, laboratory workers, and infection-control practitioners. A report by the CDC's Epidemic Intelligence Service on outbreak investigations conducted around the world from 1988 to 1999 showed that 4% of the 1099 investigations involved a potential bioterrorism agent [46]. Healthcare providers and infection-control practitioners reported 25 and 12% of the outbreaks respectively; health departments reported 31%. Reporting was delayed for up to 26 days for six outbreaks in which bioterrorism or intentional contamination was possible. Veterinarians may be the first to see evidence of bioterrorism and detection of diseases in animals may be essential to increasing the index of suspicion for human cases, because many of the bioterrorism threat agents cause zoonotic disease. Other potential sources of information leading to an outbreak investigation should include employers or organizations that may notice a high rate of illness in their workers or schools that may report a longer than usual absentee rate. Such sources can prove very valuable in countries with poor public health resources. Information from any source requires provision of a mechanism of instant reporting to the proper authorities. National health surveillance systems with further development become an important adjunct to early detection of bioterrorism events. Such systems will prove more useful for sharing information between nations and international agencies, e.g. the World Health Organization. An increasing number of public health departments are investing new surveillance systems for bioterrorism-related disease [47]. The goal of syndromic surveillance by public health departments is to detect and respond to a potential outbreak, including a potential bioterrorism event, before disease clusters are recognized clinically or specific laboratory diagnoses are made. The spectrum of activities in syndromic surveillance includes monitoring illness syndromes or events such as medication purchases that reflect the prodromes of bioterrorism-related diseases. Syndromic surveillance will be affected by the definition of syndrome categories, selection of statistical detection thresholds, selection of data sources, timelines of information management, criteria for initiating investigations, and the availability of resources for follow up. The potential of low specificity resulting in the use of resources for false alarms is one of the biggest concerns about syndromic surveillance. Syndromic surveillance systems have been deployed for continuous and event-based bioterrorism surveillance [48]. At its best, syndromic surveillance can enhance, rather than replace, traditional approaches to outbreak detection. Evaluation of detection and diagnostic decision-support systems for bioterrorism response revealed serious shortcomings and initial deficiencies [49]. The available diagnostic decision support systems will be of limited usefulness in response to a bioterrorism event.

One measure of the level of bioterrorism preparedness of public health systems is the quality of, distribution of, and compliance with disease-reporting laws [50]. It is as important that healthcare providers report unusual occurrences as it is to identify unusual patterns of disease. The Bioterrorism Preparedness and Response program at the CDC commissioned a study of local and state laws requiring the reporting of diseases caused by specific biological agents. The study of disease-reporting laws examined the reporting requirements for 24 biological agents (or disease caused by these agents) regarded as "critical biological agents." The designation was based on the potential of these agents to harm the public health if used in a terrorist attack. This commission reported significant variability in critical biological agent-reporting laws in most parts of the US as of March 31, 2001. Some jurisdictions included in this study may have revised their disease-reporting laws to include critical biological agents subsequent to the 2001 anthrax attacks. The findings of the study pointed to the need for revision and/or expansion of disease-reporting requirements to include bioterrorism-associated diseases. The authors of the report suggested that the Model State Emergency Health Powers Act (MSEHPA), a legislative template developed in 2001, may be useful to states readdressing their disease-reporting laws. The MSEHPA addresses which diseases should be reported, who should be legally required to report, the time frame and manner in which a disease should be reported, and how the disease-reporting laws should be enforced. The enactment and enforcement of such laws may provide incentives for healthcare providers to obtain necessary training and skills for diagnosing and managing potential agents of bioterrorism. As reported to the National Notifiable Disease Surveillance System (1992–1999), the incidence of disease caused by critical biological agents with the potential for bioterrorism is low in the United States [51]. Tularemia and brucellosis were the most frequently reported diseases from the group. Anthrax, plague, and equine encephalitis were rare. Each individual case report should, therefore, initially be considered a sentinel event requiring further investigation. Disease-reporting laws serve as an added reason for reporting and are a crucial element in an overall plan for bioterrorism preparedness. Several strategic legal and administrative preparations for a potential bioterrorist attack have been recommended [52]. Among these are the laws and regulations governing the confidentiality of disease-surveillance records and developing a legal and administrative procedure for sharing pertinent and relevant information with law-enforcement agencies during a bioterrorist attack.

2.5.2.4 Biosurveillance, Electronic Information, and Communications Systems

Under-reporting and significant variability in the completeness of information are major concerns with traditional disease-surveillance strategies. These delays and inconsistencies are serious challenges to public health systems in general and have the potential to impair their ability to detect or respond to a bioterrorism event. A definitive tool for biosurveillance is the electronic reporting of diagnostic results confirming the presence of a pathogen. Hoffman et al. [53] reported a multi-jurisdictional approach to biosurveillance using an electronic reporting system that ran in tandem with conventional reporting methods. More reports and more complete reports were received electronically compared with conventional methods. EMERGENCY ID NET, a surveillance system was started as a collaboration between eleven academic emergency departments and the CDC. It is an emergency department-based emerging-infections sentinel network and a useful model for the type and level of communication and cooperation necessary to detect a potential bioterrorism attack [54]. The Foodborne Disease Active Surveillance Network (FoodNet), also coordinated by the CDC, is designed to monitor and track foodborne diseases more precisely. Such programs that use standard case definitions, standardized data collection, and centralized data tracking, analysis. and reporting can be adapted, expanded, and supported to track disease syndromes potentially related to bioterrorism events. Surveillance systems based on ambulatory care settings, particularly when based on automated medical records, provide very valuable information. The Department of Defense Electronic Surveillance System for the Early Notification of Community-Based Epidemics (ESSENCE) is one of best known such systems. The system is based on encounter data from healthcare services operated by the Department of Defense. A similar system is operating in Minnesota. The use of automated ambulatory-care encounter records for detection of acute illness clusters, including potential bioterrorism events, was reported by Lazarus et al. [55]. The system is operational in eastern Massachusetts and is based on diagnoses obtained from electronic records of ambulatory-care encounters. Such systems in ambulatory-care settings complement emergency room and hospital-based surveillance by increasing the capacity for rapid identification of clusters of disease including those related to potential bioterrorism events.

2.5.2.5 Laboratory Preparedness

Rapid and accurate identification of biological agents has a pivotal role in the response to infectious disease outbreaks caused by known or hitherto unknown agents. There are approximately 174,000 laboratories operating in the US. Of these, approximately 2000 are public-health laboratories. The public-health laboratories, a network of federal, state, and local laboratories, work in undefined collaboration with private clinical laboratories. The National Laboratory Response Network (LRN) for bioterrorism became operational in the US in 1999 as part of the CDC's strategic plan for bioterrorism preparedness [56].

The CDC and the Association of Public-health laboratories developed the LRN in collaboration with the Federal Bureau of Investigation (FBI), the United State Army Medical Research Institute of Infectious Diseases (USAMRIID), the Naval Medical Research Center, and Lawrence Livermore National Laboratory. It is a multilevel laboratory system that links state and some local public-health laboratories with military, veterinary, agricultural, water, and food-testing laboratories (Appendix B). The network consists of laboratories divided into levels A, B, C, and D with progressively more stringent levels of safety, containment, and technical proficiency to perform the essential rule-out, rule-in, and referral functions required for identification of critical biological agents [57, 58]. The CDC has also established a rapid-response laboratory that will initially process samples from suspect cases, provide around-the-clock diagnostic support to bioterrorism response teams, and maintain a chain-of-custody [41]. This laboratory will also sponsor Bio-Net, modeled on Pulse-Net system, for foodborne pathogens. The system will develop and assess new rapid diagnostics before dissemination to the laboratory network, assess genetic sequencing of critical agents for establishing clonality, and monitor genotypic markers of antibiotic susceptibility and pathogenicity. Level A laboratories are hospital and other diagnostic laboratories with certified biological safety cabinets. They participate in the LRN by ruling out critical agents or referring such agents encountered in their routine work to the nearest Level B or Level C laboratory. In addition to being familiar with biosafety levels, Level A procedures, and current guidelines for handling, these laboratories must have appropriate procedures related to chain of custody, collection, preservation, and shipment of specimens. Currently these facilities mostly rely on conventional methods for ruling out specific agents as quickly as possible. The Mayo-Roche rapid anthrax test is commercially available and is a rapid-cycle real-time polymerase chain reaction [59]. This type of assay has immediate and important implications for diagnostic testing in the clinical microbiology laboratory. Because of containment requirements, most clinical laboratories are incapable of processing specimens for diagnosis of smallpox. Espy et al. reported that standard autoclaving procedures eliminated the infectivity of viruses, including smallpox, and bacteria, including *Bacillus anthracis* [60]. Target DNA was often retained for detection by LightCycler PCR. Such tests have the potential to expand the role of clinical laboratories in early detection of critical biological agents and to avoid the complexities associated with transfer of hazardous materials. Laboratories in Levels B, C, and D of the LRN have access to the biodetection assays and specialized reagents used in validated procedures for confirmation of critical agents. Level B laboratories are state and public-health laboratories with Biosafety Level 2 facilities in which Biosafety Level 3 practices are observed. Level C laboratories can perform all Level B tests in addition to the tests requiring Biosafety Level 3 containment, e.g. those involving handling of specimens containing anthrax spores. These are primarily public-health laboratories with Biosafety Level 3 facilities or certified animal facilities necessary for performing the mouse toxicity assay for botulinum toxin. Federal laboratories at the CDC and the USAMRIID are Level D laboratories with Biosafety Level 4 capacity to handle agents such as smallpox and Ebola. They also have the capacity to perform all Level B and C procedures and can identify recombinant microorganisms not recognized by conventional methods. Level D laboratories maintain an extensive culture collection of critical agents against which the isolates from a bioterrorism event may be compared in order to determine its origin. The state and territorial laboratories in the LRN are listed in Appendix B. The classification and controls for biosafety levels (BSL) 1 through 4 for microbiology laboratories are available from the CDC [61]. In addition to the role of microbiology laboratories in diagnosis and epidemiology, the field of microbial forensics is being built as a response to bioterrorism [62]. Although, in investigation of biocrimes, epidemiology and forensics are similar sciences with similar goals, the intent in microbial forensics is to identify a bioattack agent in greatest detail to establish relatedness [63]. The US government is planning a national microbial forensics system – defined as a scientific discipline dedicated to analyzing evidence from a bioterrorism act, biocrime, or inadvertent microorganism/toxin release for attribution purposes.

2.5.2.6 Biohazard Containment, Personal Protective Equipment, Decontamination and Infection-control Procedures

The best way to minimize or prevent casualties would be to detect a biological aerosol before its arrival over the target [64]. Several short-range and long-range biological detection systems are available to the US military. Although current systems are a vast improvement over capabilities available only a few years ago, they are still "detect-to-treat" systems rather than the desired "detect-to-warn" systems. They provide mostly presumptive tests for a limited number of agents.

The currently fielded protective equipment for protection against chemical agents also provides protection against biological agents. It includes the M40 protective mask, joint services lightweight integrated suit technology (JSLIST), protective gloves, multi-purpose overboots (MULO). A joint service general-purpose mask and improved gloves are under development. In addition to the threat of potential use of biological agents, military operations in tropical environments may place troops at risk of potentially lethal and contagious infectious diseases [65]. The management of such infections would be expedited by evacuating a limited number of patients to a facility with containment laboratories. The USAMRIID maintains an aeromedical isolation team. This is a rapid response team with worldwide airlift capability designed to evacuate and manage patients under high-level containment. It offers a portable containment laboratory, limited environmental decontamination, and specialized consultative expertise.

The Association of Professionals in Infection Control and Epidemiology (APIC) in cooperation with the CDC has prepared a bioterrorism readiness plan for healthcare facilities [66]. The personal protective equipment currently in use in healthcare facilities in the US is adequate for dealing with infectious agents with various modes of transmission, including the potential bioterrorism agents. De-contamination methods play an important role in the environmental control of infectious agents [64]. The methods employed must be safe for humans and animals and the materials they are used on, however. Three types of decontami-nation method can be used for bioterrorism agents:

1. Mechanical decontamination methods remove but not necessarily neutral-ize the agent. Examples include filtration of drinking water to remove water-borne pathogens, use of high-efficiency particulate air (HEPA) filters to remove aerosols of organisms, or water to wash the agent from the skin. Skin surfaces can be effectively decontaminated by careful washing with soap and water.

2. Chemical decontamination methods render biological agents harmless by affecting their viability. These disinfectants are usually in the form of a liquid, gas, or aerosol. Hypochlorite solution is a safe and tested method of chemical decontamination. For grossly contaminated skin surfaces 0.5 % sodium hypochlorite solution with a contact time of 10–15 min is recom-mended. The standard stock Clorox is 5.25 % sodium hypochlorite solution. Adding one part Clorox to nine parts water (1:9) makes a 0.5 % solution which should be prepared fresh to keep the pH in the alkaline range. Use of chlorine solution is contraindicated for open body cavity wounds, because of the risk of adhesion formation and of brain and spinal-cord injuries. A 5 % hypochlorite solution should be used for decontamination of fabric clothing or equipment. A contact time of 30 min is needed for equipment and should be followed by thorough rinsing and oiling of the metal surfaces to prevent corrosion.

3. Physical decontamination is a physical means of rendering biological agents harmless; for example heat and radiation can be used to decontaminate objects. Dry heat at 160 °C for 2 h and autoclaving with steam at 121 °C and 15 psig pressure for 20 to 30 min are used to sterilize objects. Solar ultra-violet (UV) radiation in combination with drying has a disinfectant effect and, with oxidation, can be relied upon for natural inactivation of biological agents in the outdoor environment. Biological agents deposited on the soil would be subject to degradation by environment stressors and competing soil microflora. On the basis of simulant studies secondary re-aerosolization is not considered a human health hazard. Re-aerosolization can be mini-mized by the use of a dust-binding spray if grossly contaminated terrain, streets, or roads must be passed. If needed, chlorine-calcium or lye can be used for environmental decontamination.

Standard precautions are employed routinely in the care of all patients. Potential agents of bioterrorism that can be transmitted person-to-person by the respiratory route require the most stringent patient-isolation precautions in the healthcare setting. The isolation precautions needed for patients with infectious diseases, including those caused by potential bioterrorism agents, are categorized on the basis of the most likely mode of transmission (Appendix C).

2.5.2.7 Pharmaceutical Readiness

Because diseases caused by many potential bioterrorism agents are rare and fatal, it is virtually impossible to conduct rigorous, controlled studies in humans to evaluate the safety and efficiency of antimicrobial agents [67]. The off-label use of therapeutic agents may become necessary but should be minimized and used in consultation with infectious-disease experts and bodies like the CDC and the Food and Drug Administration (FDA). The President of the United States issued Executive Order 13139 on September 30, 1999, which outlines the conditions under which off-label pharmaceuticals and investigational new drugs (IND) can be administered to US service men [64]. The executive order provides the Secretary of Defense guidance regarding the provision and use of off-label or IND products for their intended use as antidotes to chemical, biological, or radiological weapons. It provides the circumstances and controls under which IND products may be used. Informed consent must be obtained from the service member unless it is not feasible, is contrary to the best interests of service members, or obtaining informed consent is not in the best interest of national security.

The United States Food and Drug Administration (FDA) is part of an inter-agency group preparing for response in a civilian emergency [68]. The group represents the Department of Defense, the Veteran's Administration, Centers for Disease Control and Prevention (CDC), National Institutes of Health (NIH), and the Office of Emergency Response. The regulations regarding approval of therapeutic agents by the FDA are based on science, law, and public-health considerations. Research on the design of new vaccines, pathogens, and mechanisms of replication of biological warfare agents and standards to expedite the review of new active and passive immunization products is conducted at the FDA's Center for Biologics Evaluation and Research. The FDA has also proposed standards for use, when appropriate, of animal efficacy data as a surrogate for clinical trials. The manufacturers will still be expected to provide conventional data to demonstrate the safety of any agent and immunogenicity of candidate vaccines. The three major types of pharmacological intervention after a bioterrorism event will be:

1. Antimicrobial treatment and prophylaxis. Although treatment of a suspected or confirmed infection by a specific antimicrobial agent will be based on known patterns of antimicrobial susceptibility, the possibility of genetically engineered organisms with possible resistance to otherwise effective agents will need to be considered. The agents selected for treatment can also be used for mass chemoprophylaxis except that only oral rather than intravenous preparations of antimicrobial agents will be practical. The antimicrobial agents of choice for bioterrorism related agents for which treatment is available have been discussed and described extensively [69–72]. The use of antibiotics for mass protection has the inherent disadvantages of side

effects, intolerances, noncompliance, and development of resistance and lack of efficacy.

2. Vaccines. Active immunization by vaccines is the optimum way of protecting masses against infectious agents, including potential agents of bioterrorism. Because these diseases are no longer seen in the US population, routine immunization against potential agents of bioterrorism is not practiced except for military personnel. The emergency personnel, including the healthcare providers, are a target group for vaccination because of high risk of infection, potential transmission to other groups, and their essential role in managing public health emergencies. Currently vaccines for human use are available for anthrax and smallpox. Anthrax vaccines with better efficacy and fewer injections than for the current vaccine are under evaluation. Smallpox immunization of healthcare workers to create "smallpox-response teams" was started in the US in December 2002. The complexities, difficulties, and current status of this initiative are fully described in Chapter 10. Major medical concerns are symptoms and side-effects suggestive of systemic involvement [73]. Smallpox vaccination can be protective even when given 3 to 4 days after infection (exposure). Therefore, a vaccination program can be started as soon as a potential event involving smallpox virus is suspected. Vaccination can circumvent the need for antimicrobial prophylaxis or reduce the duration for which it would otherwise be used.

3. Passive antibody to provide immediate immunity. Passive (preformed) antibody therapy has low toxicity, high specific, and immediate activity and does not depend on the host's ability to mount an immune response. Antitoxin (antibody) against *Clostridium botulinum* toxin is currently used as the treatment of choice for botulism. Specific antibodies are known to be active against the major agents of bioterrorism including anthrax, smallpox, botulinum toxin, tularemia, and plague [74]. A neutralizing antibody preparation can be generated much faster than new biological agents can be developed. Historically, the 1905 epidemic of meningococcal meningitis in New York City serves as an example. An effective horse serum was generated within months and used to treat patients before the epidemic abated naturally. Use of antibody-based therapy in clinical practice is limited by factors such as cost, need for a specific diagnosis, and lack of efficacy for established infections. Passive immunization is more effective for prevention than for treatment. As discussed in the chapter on anthrax (Chapter 5), the usefulness of passive antibody therapy can be enhanced by developing monoclonal antibodies with capability to neutralize virulence factors of microbial agents. The development of antibody-based therapies for infectious diseases has been hindered by the cost, small market size, and the availability of many antimicrobial drugs in the recent years. Because, in biodefense initiatives, the market size equals the potentially vulnerable population, and because of the need to replenish stockpiles, the economic outlook may be more attractive to industry. Antibody-based agents would provide a means of conferring immediate immunity on susceptible persons

and substantially reducing the threat of many biological agents. This intervention would also provide additional time for immunization against the agents for which vaccines are available.

Stockpiles of supplies, antimicrobial agents, and vaccines have been created by the Department of Health and Human Services and maintained by the CDC [19], the national repository of pharmaceuticals and medical material. The National Pharmaceutical Stockpile (NPS) is bundled into "push-packs" that can be deployed, by commercial cargo, to the scene of a biological or chemical weapons attack within 12 h of request by a state. Vendor-managed inventory (VMI) is the follow up component of the NPS. Antimicrobial agents from pharmaceutical manufacturers are intended to be shipped to the site of a bioterrorism event within 24–36 h of a request.

2.5.3
Political Preparedness

The single most important issue in any emergency and preparedness plan is to have an established chain of command. The question of who is "in charge" must be adequately answered before an emergency. Although the Department of Defense has the greatest capability for defense against biological agents, responsibility for dealing with public-health effects of biological agents falls on many bodies including federal, state, and local government and, ultimately, the civilian medical community. All healthcare facilities, public-health organizations and government agencies should have an action plan with responsible personnel and their deputies in place before the incident.

In the aftermath of Sarin gas attacks in Japan and the Oklahoma City bombing, President Clinton signed Presidential Decision Directive 29 (PPD-39) which addressed how the US would deal with a potential attack with weapons of mass destruction. The US Congress followed this directive with the Defense Against Weapons of Mass Destruction Act. PPD 39 designates the Federal Bureau of Investigation (FBI) as the lead agency for the crises plan and the Federal Emergency Management Agency (FEMA) is charged with ensuring that the federal response to the consequences of bioterrorism is adequate [56]. At the request of a State Health Agency the CDC will deploy response teams to investigate unexplained illnesses or unusual etiologic agents and provide onsite consultations for disease control and medical management. The National Pharmaceutical Stockpile (NPS) maintained by the CDC will be deployed with investigative expertise.

The Office of Homeland Security (OHS) was created with executive order (E. O. 13228) by President Bush in the aftermath of the September 11, 2001 attacks. The president's proposal to upgrade OHS to a Department of Homeland Security was approved by the House on July 26, 2002. The legislation mandating the Department was signed into law on November 25, 2002. Soon after the Department became operational, a Presidential Directive designated the Secretary as the principal federal official for domestic incidents management. The Homeland Security

Act contains several provisions requiring reports to Congress by the Secretary of Homeland Security, Department of Homeland Security Officials, other executive branch officials, and the General Accounting office.

2.5.4
Bioterrorism Preparedness – Global Avenues

The threats posed by emerging and re-emerging infectious diseases and the resistance of many microbial agents to currently available drugs have serious global implications. With the level of interconnection in the world today, the spread of infectious diseases is easier than ever. Most countries where infectious diseases are still the most common cause of morbidity and mortality do not have the public health infrastructure or the resources to prepare against bioterrorism events or the spread of bioterrorism agents. The World Health Organization, an international health agency has over 50 years of experience providing response to outbreaks and preventing international spread of disease [12]. WHO has permanently positioned geographical resources, including the Geneva Headquarters, six regional offices, and 141 country offices located in close proximity to ministries of health. The country offices are concentrated in areas where epidemics are frequent and the likelihood of emergence of new diseases is high. All country offices are linked electronically to WHO and its global network of institutional resources and collaborators. Each country office is staffed with medical experts and epidemiologists and has essential equipment and resources needed for prompt outbreak investigation. The offices also facilitate the arrival and functioning of international assistance. The WHO has 250 laboratories and institutions as its designated collaborating centers. It has a very close working relationship with the CDC, including direct secondment of staff. In addition, technical support to the WHO is provided by US agencies such as USAID and many laboratories included in the US Department of Defense Global Emerging Infections Surveillance and Response system and their counterparts in other WHO member states. Other US agencies that provide technical and financial support to WHO include the National Institutes of Health (NIH) through its Fogarty International Center, National Center for Environmental Prediction, NASA–Goddard space flight Center, and the State Department's Bureau of Population, Refugees, and Migration (PRM). The recent strengthening of the CDC and bioterrorism preparedness efforts by the US has benefited WHO and a large number of countries weakened by repeated outbreaks and epidemics.

WHO is politically neutral and often the only or most important source of authoritative advice and technical assistance pertaining to outbreaks of infectious diseases in many parts of the world. With its deep experience in coordinating field operations needed to control infectious diseases, it would have the lead role of investigating bioterrorism related events. The epidemiological techniques and laboratory support needed to investigate natural and deliberate outbreaks are the same. WHO coordinates a large number of electronic "detective" systems and data bases within its surveillance networks. These networks operate in real time and keep watch over known risks and unexpected or unusual disease events. FluNet, established to monitor global influenza virus activity over 50 years ago, has served as the prototype for many subsequent systems. In recent years, WHO has added innovative mechanisms to respond to previously unknown diseases and unexpected or unusual disease events. It uses a semi-automatic electronic system developed by Health Canada to scan the world by continuously and systematically crawling websites, news wires, public health e-mail services, and electronic discussion groups, including the US based Pro-MED and local online newspapers. Suspicious reports are investigated each morning by a WHO team responsible for outbreak verification. This is followed by the use of WHO's technical and geographical resources to verify the presence of an outbreak, as and when appropriate. The World Health Assembly, the supreme governing body of WHO, adopted by consensus a resolution on global health security in May, 2001. This resolution has considerably strengthened WHO's capacity to act and the organization is now in a position to investigate and verify rumored outbreaks even before it receives an official notification from the concerned government. This heightened vigilance has been combined with rapid response by formation of the Global Outbreak Alert and Response Network by WHO. Because international assistance can involve many agents from many nations, the coordination is facilitated by operational procedures by WHO. Guidelines for the behavior of foreign nationals during and after field operations in the host country have also been issued.

For obvious reasons, building and strengthening national epidemic detection and response is far more efficient and cost-effective than mounting an international response. WHO helps countries strengthen their epidemiological and laboratory capacity. Examples of these efforts include the Training Programs in Epidemiology and Public Health Interventions Network (TEPHINET) in collaboration with the CDC and a working group on long-term preparedness for outbreak response. Consequently, early warning and response networks (EWARN) have been started in partnership with nongovernmental organizations in the field. Any local and national effort related to public-health functions can only succeed by sustained improvements in civil administration. Simple measures like establishment of a community registry of births and deaths and a community-based early-warning surveillance and response system for priority infectious diseases in Uganda have led to quick detection of cases of hemorrhagic fever by local staff. Local vigilance and strengthened national capacity make it possible to defend global health security. WHO guidance on public health response to biological and chemical weapons and laboratory safety are provided in recently updated manuals [75, 76].

Global solutions to infectious disease threats should ideally be aimed at prevention. The global eradication of diseases like smallpox and the significant decrease in mortality from many other infectious diseases attest to the role that optimum use of current vaccines and future vaccines have in controlling diseases and making the population immune to diseases including those caused by potential bioterrorism agents. Vaccine development and supplies have not kept pace with need. Because of multiple factors including the time and resources needed for vaccine development, low profitability, and potential litigation for alleged vaccine-induced injury, many companies have been driven out of the vaccine market [77]. Vaccine supplies fall short from time to time even in the US. Global support and strengthening of vaccine manufacture and equitable distribution is needed to control diseases currently preventable by use of vaccines and bring better and more vaccines to the consumer.

International efforts to protect global health are governed by laws, as are national efforts. These include the humanitarian laws that govern war and its aftermath, international laws and conventions (e.g. BWC) to counteract various types of threats, and the international health laws [78–80]. International governance in infectious diseases started with the 1851 International Sanitary Conference. During the early years, three horizontal international legal regimes related to infectious diseases appeared – the classical, organizational, and trade regimes. The trade regime created in the 1851–1951 period is represented by the General Agreement on Tariffs and Trade (GATT, 1947). GATT liberalized trade but recognized that states may restrict trade to protect public health. In the organizational legal regime, international health organizations were created to deal with infectious diseases and other public health problems. WHO is the leading representative of this governance framework. International Sanitary Conventions adopted until World War II and the International Sanitary Regulations (1951) represent the classical regime. The purpose of International Sanitary Regulations, later renamed the International Health Regulations (IHR), is "to ensure the maximum protection against the international spread of disease with minimum interference with world traffic." The IHR addressed only three diseases (cholera, yellow fever, and plague) and authorized WHO to use information provided only by member states. Given the current natural infectious disease scene and the threat of bioterrorism, these regulations were considered outdated and ineffective. The revised International Health Regulations (IHR) were adapted by the World Health Assembly in 2005 and are discussed in Chapter 11.

References

1 Drexler, M. **2002** Secret Agents: the menace of emerging infections. Joseph Henry Press.

2 Molyneux, D. **2004** "Neglected" diseases but unrecognized successes – challenges and opportunities for infectious disease control. *Lancet* 364:380–83.

3 Osterholm, M **2005** Preparing for the next pandemic. *N Engl J Med* 352:1839–1842.

4 Hien, T. T., Jong, M. D., Farrar, J. **2004** Avian influenza – A challenge to global health care structures. *N Engl J Med* 351; 23:2363–2365.

5 Centers for Disease Control and Prevention. **1998** Preventing Emerging infectious diseases: A strategy for the 21st century. MMWR Vol. 47/ No. RR-15

6 Simons, L. M. **2002** Weapons of mass destruction. *National Geographic* 2–35.

7 The CQ Researcher: Chemical and biological weapons. **1997**The CQ Press, Vol. 7, No. 4:85.

8 Christopher, G. W., Gieslak, T. J., Pavlin, J. A. et al. **1997** Biological warfare. A historical perspective. *JAMA* 278:412–417.

9 Wheelis, M. **2000** Investigating disease outbreaks under a protocol to the biological and toxin weapons convention. *Emerg Inf Dis* 66:595–600.

10 Dorey, E. **2001** US rejects stronger bioweapons treaty. *Nature Biotechnology.* 19:793.

11 Cherry, C. L., Kainer, M. A. and Ruff. T. A. **2003** Biological weapons preparedness: the role of physicians. *Int Med J* 33:242–253.

12 Heymann, D. L. **2001** Strengthening global preparedness for defense against infectious disease threats. Committee on Foreign Relations, United States Senate. Hearing on the *Threat of bioterrorism and the spread of infectious diseases.* 1st Session, 107th Congress.

13 Editorial. **2002** Obstacles to biodefense *Nature* 419:1.

14 County public health preparedness. **2002** *NACo* www.naco.org/pubs/ surveys/pubhealth/index.cfm

15 Hoard, M. L. Williams, J. M., Helmkamp, J. C., et al. **2002** Preparing at the local level for events involving weapons of mass destruction. *Emerg Inf Dis* 8:1006–1007.

16 DiGiovanni, C., Reynolds, B., Harwell, R., et al. **2003** Community reaction to bioterrorism: Prospective study of simulated outbreak. *Emerg Inf Dis* 9:708–712

17 Boulton, M. D., Abellera, J. Lemmings. J., et al. **2005** Terrorism and emergency preparedness in state and territorial public health departments – United States, 2004. *MMWR* Vol. 54/No. 18:459–460.

18 Commentary. **2000** Lessons learned from a full-scale bioterrorism exercise. *Emerg Inf Dis* 6:652–654.

19 Inglesby, T. V., Grossman, R. and O'Toole, R. **2001** A plague on your city: Observations from TOPOFF. *Clin Inf Dis* 32:436–444.

20 O'Toole, T., Mair, M. and Inglesby, T. V. **2002** Shining light on "Dark Winter". *Clin Inf Dis* 34:972–983.

21 Atlas, R. M. **2002** Bioterrorism: From threat to reality. *Annu Rev Microbiol* 56:167–185.

22 Frase-Blunt, M. **2003** "Operation TOPOFF 2" Bioterrorism exercise offers educational lessons. AAMC Reporter. www.aamc/org/august03/bioterrorism.html

23 Nelson, R. **2005** Simulation shows lack of readiness for bioterrorism attack. *Lancet Infectious Disease* 5:139.

24 Gilsdorf, J. R. and Zilinskas, R. A. **2005** New considerations in infectious disease outbreaks: The threat of genetically modified microbes. *Clin Inf Dis* 40:1160– 1165.

25 Fraser, C. M. and Dando, M. R. **2001** Genomics and future biological weapons: the need for preventive action by the biomedical community. *Nature Genetics* 29:253–256.

26 Dennis, C. **2001** The bugs of war. *Nature* 411:232–235.

27 Caudle, L. D. **1997** The biological warfare threat. In: Zajtichuk R., ed. Textbook of military medicine. Office of The Surgeon General, Department of the Army, Washington, D. C.451–466.

28 Alibek, K. and Handleman, S. **1999** Biohazard: the chilling true story of the largest covert biological weapons program in the world – told from inside by the man who ran it. Random House, New York, New York.

29 Popov, S. **2000** Interview – Serguei Popov J. Homeland Security. Available at http://www.homelandsecurity.org/journal/Interviews/PopopvInterview_001107.htm.

30 National Research Council. **2004** Biotechnology research in an age of terrorism. The National Academies Press, Washington, D. C.

31 Centers for Disease Control and Prevention, U. S. Department of Health and Human Services, **2001** The Public Health Response to Biological and Chemical Terrorism, Interim Planning Guidance for State Public Health Officials available at http://www.be.cdc.gov/Documents/Planning/PlanningGuidance. PDF

32 Gostin, L. O., Sapsin, J. W., Teret, S. P., et al. **2002** The model state emergency health powers act: planning for and response to bioterrorism and Naturally occurring infectious diseases. *JAMA* 288:622–628.

33 Kaufman, A. F., Meltzer, M. I. and Schmid, G. P. **1997** The economic impact of a bioterrorist attack: Are prevention and postattack intervention program justifiable. *Emerg Inf Dis* 3:83–93.

34 Annas, G. J. **2002** Bioterrorism, public health, and civil liberties. *N Engl J Med* 346:1337–1342.

35 Federal Register, 42 CFR 72. **2005** Department of Health and Human Services.

36 Centers for Disease Control and Prevention **2003** Select Agent Program. FAQ for New Regulation. CDC: Office of the Director:www.cdc.gov/od/sap/addres/htm.

37 Siegrist, D. W. **1999** The threat of biological attack: Why concern now. *Emerg Inf Dis*, 5:505–508.

38 Army Reserve National Guard. **1998** Stakeholders III Conference Medical Panel. National Guard Bureau, Arlington, Virginia.

39 Tonat, K. **1999** Office of Emergency Preparedness, U. S. Department of Health and Human Services. Panel discussion at conference "Integrating Medical and Emergency Response," Washington, D. C.

40 Mitchel, A. **1999** Office of State and Local Domestic Preparedness Support, U. S. Department of Justice. Panel discussion at conference "Integrating Medical and Emergency Response," Washington, D. C.

41 Khan, A. S., Morse, S. and Lillibridge, S. **2000** Public-health preparedness for biological terrorism in the USA. *Lancet* 356:1179–1182.

42 Centers for Disease Control and Prevention **2005** Assessment of epidemiologic capacity in state and territorial health departments – United States, 2004. MMWR Vol. 54/ No. 18:457–458.

43 General Accounting Office. **2000** Combating terrorism: need to eliminate duplicate federal weapons of mass destruction training [GAO/NSAID-00–64].

44 General Accounting Office. **1998** Combating terrorism: observations on the Nunn– Lugar–Domenici Domestic Preparedness Program. [GAO/T-NSAID-99–16].

45 Hupert, N., Chege, W., Bearman, G., et al. **2004** Antibiotics for anthrax. *Arch Intern Med*; 164:2012–2016.

46 Ashford, D. A., Kaiser, R. M., Bales, M. E., et al. **2003** Planning against biological terrorism: Lessons from outbreak investigations. *Emerg Inf Dis* 9:515–519.

47 Buehler, J. W., Berkelman, Hartley, D. M. and Peters, C. J. **2003** Syndromic surveillance and bioterrorism-related epidemics. *Emerg Inf Dis* 9:1197–1203.

48 Bravata, D. M., McDonald, K. M., Smith, W. M., et al. **2004** Systematic review: Surveillance systems for early detection of bioterrorism-related diseases. *Ann Intern Med.*; 140:910–922.

49 Bravata, D. M., Sundaram, V., McDonald, K. M., et al. **2004** Evaluating detection and diagnostic decision support systems for bioterrorism response. *Emerg Inf Dis.* 10:100–107.

50 Horton, H. H., Misrahi, J. J., Matthew, G. W., et al. **2002** Critical biological agents: Disease reporting as a tool for determining bioterrorism preparedness. *Journal of Law, Medicine & Ethics,* 30:262–266.

51 Chang, M., Glynn, K. and Groseclose, S. L. **2003** Endemic, notifiable bioterrorism-related diseases, United States, 1992 – 1999. *Emerg Inf Dis* 9:556–563.

52 Hoffman, R. E. **2003** Preparing for a bioterrorist attack: Legal and administrative strategies. *Emerg Inf Dis* 9:241–245.

53 Hoffman, M. A., Wilkinson, T. H., Bush, A., et al. **2003** Multijurisdictional approach to biosurveillance, Kansas City. *Emerg Inf Dis* 9:1281–1285.

54 Richards, C. F., Burstein, J. L., Waeckerle, J. F., et al. **1999** Emergency physicians and biological terrorism. *Ann of Emerg Med* 34:1 – 11.

55 Lazarus, R., Kleinman, K. Dashevsky, I., et al. **2002** Use of automated ambulatory-care encounter records for detection of acute illness clusters, including potential bioterrorism events. *Emerg Inf Dis* 8:753–760.

56 Centers for Disease Control and Prevention. **2000** MMWR Biological and chemical terrorism: Strategic plan for preparedness and response. Vol. 49/ No. RR-4.

57 Meyer, R. F. **2002** Bioterrorism preparedness for the public health and medical communities. *Mayo Clin Proc.* 77:619–621.

58 Khardori, N. **2004** Preparedness for bioterrorism. In: Meyers, R. A. (Ed.) *Encyclopedia of Molecular cell biology and molecular medicine.*, Vol. 10. Wiley–VCH, Weinheim, pp. 507–528.

59 Uhl, J. R., Bell, C. A., Sloan, L. M., et al. **2002** Application of rapid-cycle real-time polymerase chain reaction for the detection of microbial pathogens: The Mayo– Roche rapid anthrax test. *Mayo Clin Proc.* 77:673–680.

60 Espy, M. J., Uhl, J. R., Sloan, L. M., et al. **2002** Detection of Vaccinia virus, Herpes Simples virus, Varicella–Zoster virus, and *Bacillus anthracis* DNA by LightCycler Polymerase chain reaction after autoclaving: Implications for biosafety of bioterrorism agents. *Mayo Clin Proc* 77:624–628

61 LRN Network **2004** http://www.cdc.gov/od/ohs/pdffiles/Module%202%20-%Biosafety.pdf and http://www.cdc.gov/od/ohs/biosfty/bmbl4bmbl4toc.htm

62 Budowle, B. Schutzer, S. E., Einsein, A., et al. **2003** Building microbial forensics as a response to bioterrorism. *Science* 301:1852–1853.

63 Keim, P.. **2003** Microbial Forensics: A scientific assessment. Report from a colloquium sponsored by the American Academy of Microbiology. Burlington, Vermont.

64 USAMRIID's Medical Management of biological casualties handbook. **2004** 5th Edition U. S. Army Medical Research Institute of Infectious Diseases Fort Detrick, Frederick, Maryland.

65 Christopher, G. W. and Eitzen, E. M. **1999** Air evacuation under high-level biosafety containment: The aeromedical isolation team. *Emerg Inf Dis* 4:241–246.

66 English, J. F., Cundiff, M. Y., Malone, J. D., et. al.**1999** Bioterrorism readiness plan: A template for healthcare facilities. APIC Bioterrorism Task Force and the CDC Hospital Infections Program Bioterrorism Working Group. 1 – 33.

67 McKinney, W. P., Bia, F. J., Stewart, C. S. et al. **2001** "Bioterrorism: An update for clinicians, pharmacists, and emergency management planners. Emergency medicine consensus reports. *American Health Consultants.* Atlanta, Georgia.

68 Zoon, K. C. **1999** Vaccines, pharmaceutical products and bioterrorism: Challenges for the U. S. Food and Drug Administration. *Emerg Inf Dis* 5:534

69 Inglesby, T. V. O'Toole, T., Henderson, D.A, et al. **2002** *Anthrax as a biological weapon – Updated recommendations for management* in Bioterrorism – Guidelines for medical and public health management. Edited by Henderson, D. A., Inglesby, T. V. and O'Toole, T. American Medical Association. Chicago, Illinois.

70 Inglesby, T. V., Dennis, D. T. Henderson, D. A., et al. **2002** Plague as a biological weapon in Bioterrorism – Guidelines for medical and public health management. Edited by Henderson, D. A., Inglesby, T. V. and O'Toole, T. American Medical Association. Chicago, Illinois.

71 Arnon, S. S., Schecter, R., Inglesby, T. V., et al. **2002** Botulinum toxin as a biological weapon. In bioterrorism – Guidelines for medical and public health management. Edited by Henderson, D. A., Inglesby, T. V. and O'Toole, T. American Medical Association. Chicago, Illinois.

72 Dennis, D. T., Inglesby, T. V., Henderson, D. A., et. al.**2002** Tularemia as a biological weapon. In Bioterrorism – Guidelines for medical and public health management. Edited by Henderson, D. A., Inglesby, T. V. and O'Toole, T. American Medical Association. Chicago, Illinois.

73 Baggs, J., Chen, R. T., Damon, I. K. , et al. **2005** Safety profile of smallpox vaccine: Insights from the laboratory worker smallpox vaccination program. *Clin Infect Dis* 40:1133–1140.

74 Casadevall, A. **2002** Passive antibody administration (Immediate Immunity) as a specific defense against biological weapons. *Emerg Inf Dis* 8:833–884.

75 World Health Organization **2004** Public health response to biological and chemical weapons. WHO guidance. 1–340.

76 World Health Organization **2004** laboratory biosafety manual, 3rd Edition 1–278.

77 Cohen, J. **2002** U. S. Vaccine supply falls seriously short. *Science* 295:1998–2005.

78 Editorial. **2003** Laws, war, and public health. *The Lancet* 361:1399.

79 Holdstock, D. **2001** Reacting to terrorism. The response should be through law not war. *BMJ* 323:822.

80 Fidler, D. P. **2003** Emerging trends in international law concerning global infectious disease control. *Emerg Inf Dis* 9:285–290.

Appendix A:

HHS and USDA Select Agents and Toxins – 7 CFR Part 331, 9 CFR Part 121, and 42 CFR Part 73

HHS Select agents and toxins

Abrin
Cercopithecine herpesvirus 1 (Herpes B virus)
Coccidioides posadasii
Conotoxins
Crimean-Congo hemorrhagic fever virus
Diacetoxyscirpenol
Ebola viruses
Lassa fever virus
Marburg virus
Monkeypox virus
Ricin
Rickettsia prowazekii
Rickettsia rickettsii
Saxitoxin
Shiga-like ribosome inactivating proteins
South American Haemorrhagic Fever viruses
    Flexal
    Guanarito
    Junin
    Machupo
    Sabia
Tetrodotoxin
Tick-borne encephalitis complex (flavi) viruses
    Central European Tick-borne encephalitis
    Far Eastern Tick-borne encephalitis
    Kyasanur Forest Disease
    Omsk Hemorrhagic Fever
    Russian Spring and Summer encephalitis
Variola major virus (Smallpox virus)
Variola minor virus (Alastrim)
Yersinia pestis

Overlap select agents and toxins

Bacillus anthracis
Botulinum neurotoxins
Botulinum neurotoxin producing species of
Clostridium
Brucella abortus
Brucella melitensis
Brucella suis
Burkholderia mallei (formerly *Pseudomonas mallei*)
Burkholderia pseudomallei (formerly *Pseudomonas pseudomallei*)
Clostridium perfringens epsilon toxin
Coccidioides immitis

USDA Select agents and toxins

African horse sickness virus
African swine fever virus
Akabane virus
Avian influenza virus (highly pathogenic)
Bluetongue virus (Exotic)
Bovine spongiform encephalopathy agent
Camel pox virus
Classical swine fever virus
Cowdria ruminantium (Heartwater)
Foot-and-mouth disease virus
Goat pox virus
Japanese encephalitis virus
Lumpy skin disease virus
Malignant catarrhal fever virus
(Alcelaphine herpesvirus type 1)
Menangle virus
Mycoplasma capricolum/ *M.F38*/*M. mycoides capri* (contagious caprine pleuropneumonia)
Mycoplasma mycoides mycoides (contagious bovine pleuropneumonia)
Newcastle disease virus (velogenic)
Peste des petits ruminants virus
Rinderpest virus
Sheep pox virus
Swine vesicular disease virus
Vesicular stomatitis virus (Exotic)

USDA Plant protection and quarantine (PPQ) select agents and toxins

Candidatus Liberobacter africanus
Candidatus Liberobacter asiaticus
Peronosclerospora philippinensis
Ralstonia solanacearum race 3, biovar 2
Schlerophthora rayssiae var *zeae*
Synchytrium endobioticum
Xanthomonas oryzae pv. *oryzicola*
Xylella fastidiosa (citrus variegated chlorosis strain)

HHS Select agents and toxins [CONTD]
Coxiella burnetii
Eastern Equine Encephalitis virus
Francisella tularensis
Hendra virus
Nipah virus
Rift Valley fever virus
Shigatoxin
Staphylococcal enterotoxins
T-2 toxin
Venezuelan Equine Encephalitis virus

Appendix B:
Emergency Response Contacts in the United States

State and Territorial Public Health Directors (Listed by State)
Reproduced with permission from USAMRIID's Medical Management of Biological
Casualties Handbook 2004 [64].

Alabama
Department of Public Health
State Health Officer
Phone No. (334) 206–5200
Fax No. (334) 206–2008

Alaska
Division of Public Health
Alaska Department of Health and Social Services
Director
Phone No. (907) 465–3090
Fax No. (907) 586–1877

American Samoa
Department of Health
American Samoa Government
Director
Phone No. (684) 633–4606
Fax No. (684) 633–5379

Arizona
Arizona Department of Health Services
Director
Phone No. (602) 542–1025
Fax No. (602) 542–1062

Arkansas
Arkansas Department of Health
Director
Phone No. (501) 661–2417
Fax No. (501) 671–1450

California
California Department of Health Services
State Health Officer
Phone No. (916) 657–1493
Fax No. (916) 657–3089

Colorado
Colorado Department of Public Health & Environment
Executive Director
Phone No. (303) 692–2011
Fax No. (303) 691–7702

Connecticut
Connecticut Department of Public Health
Commissioner
Phone No. (860) 509–7101
Fax No. (860) 509–7111

Delaware
Division of Public Health
Delaware Department of Health and Social Services
Director
Phone No. (302) 739–4700
Fax No. (302) 739–6659

District of Columbia
DC Department of Health
Acting Director
Phone No. (202) 645–5556
Fax No. (202) 645–0526

Florida
Florida Department of Health
Secretary and State Health Officer
Phone No. (850) 487–2945
Fax No. (850) 487–3729

Georgia
Division of Public Health
Georgia Department of Human Resources
Director
Phone No. (404) 657–2700
Fax No. (404) 657–2715

Guam
Department of Public Health & Social Services
Government of Guam
Director of Health
Phone No. (67 l) 735–7102
Fax No. (671) 734–5910

Hawaii
Hawaii Department of Health
Director
Phone No. (808) 586–4410
Fax No. (808) 586–4444

Idaho
Division of Health
Idaho Department of Health and Welfare
Administrator
Phone No. (208) 334–5945
Fax No. (208) 334–6581

Illinois
Illinois Department of Public Health
Director of Public Health
Phone No. (217) 782–4977
Fax No. (217) 782–3987

Indiana
Indiana State Department of Health
State Health Commissioner
Phone No. (317) 233–7400
Fax No. (317) 233–7387

Iowa
Iowa Department of Public Health
Director of Public Health
Phone No. (515) 281–5605
Fax No. (515) 281–4958

Kansas
Kansas Department of Health and Environment
Director of Health
Phone No. (785) 296–1343
Fax No. (785) 296–1562

Kentucky
Kentucky Department for Public Health
Commissioner
Phone No. (502) 564–3970
Fax No. (502) 564–6533

Louisiana
Louisiana Department of Health and Hospitals
Asst Secretary and State Health Officer
Phone No. (504) 342–8093
Fax No. (504) 342–8098

Maine
Maine Bureau of Health
Maine Department of Human Services
Director
Phone No. (207) 287–3201
Fax No. (207) 287–4631

Mariana Islands
Department of Public Health & Environmental Services
Commonwealth of the Northern Mariana Islands
Secretary of Health and Environmental Services
Phone No. (670) 234–8950
Fax No. (670) 234–8930

Marshall Islands
Republic of the Marshall Islands
Majuro Hospital
Minister of Health & Environmental Services
Phone No. (692) 625–3355
Fax No. (692) 625–3432

Maryland
Maryland Dept of Health and Mental Hygiene
Secretary
Phone No. (410) 767–6505
Fax No. (410) 767–6489

Massachusetts
Massachusetts Department of Public Health
Commissioner
Phone No. (617) 624–5200
Fax No. (617) 624–5206

Michigan
Community Public Health Agency
Michigan Department of Community Health
Chief Executive and Medical Officer
Phone No. (517) 335–8024
Fax No. (517) 335–9476

Micronesia
Department of Health Services
FSM National Government
Secretary of Health
Phone No. (691) 320–2619
Fax No. (691) 320–5263

Minnesota
Minnesota Department of Health
Commissioner of Health
Phone No. (651) 296–8401
Fax No. (651) 215–5801

Mississippi
Mississippi State Department of Health
State Health Officer and Chief Executive
Phone No. (601) 960–7634
Fax No. (601) 960–7931

Missouri
Missouri Department of Health
Director
Phone No. (573) 751–6001
Fax No. (573) 751–6041

Montana
Montana Department of Public Health & Human Services
Director
Phone No. (406) 444–5622
Fax No. (406) 444–1970

Nebraska
Nebraska Health and Human Services System
Chief Medical Officer
Phone No. (402) 471–8399
Fax No. (402) 471–9449

Nevada
Division of Health
Nevada State Department of Human Resources
State Health Officer
Phone No. (702) 687–3786
Fax No. (702) 687–3859

New Hampshire
New Hampshire Department of Health & Human Services
Medical Director
Phone No. (603) 271–4372
Fax No. (603) 271–4827

New Jersey
New Jersey Department of Health & Senior Services
Commissioner of Health
Phone No. (609) 292–7837
Fax No. (609) 292–0053

New Mexico
New Mexico Department of Health
Secretary
Phone No. (505) 827–2613
Fax No. (505) 827–2530

New York
New York State Department of Health
ESP-Corning Tower, 14th Floor
Albany, NY 12237
Commissioner of Health
Phone No. (518) 474–2011
Fax No. (518) 474–5450

North Carolina
NC Department of Health and Human Services
State Health Director
Phone No. (919) 733–4392
Fax No. (919) 715–4645

North Dakota
North Dakota Department of Health
State Health Officer
Phone No. (701) 328–2372
Fax No. (701) 328–4727

Ohio
Ohio Department of Health
Director of Health
Phone No. (614) 466–2253
Fax No. (614) 644–0085

Oklahoma
Oklahoma State Department of Health
Commissioner of Health
Phone No. (405) 271–4200
Fax No. (405) 271–3431

Oregon
Oregon Health Division
Oregon Department of Human Resources
Administrator
Phone No. (503) 731–4000
Fax No. (503) 731–4078

Palau, Republic of
Ministry of Health
Republic of Palau
Minister of Health
Phone No. (680) 488–2813
Fax No. (680) 488–1211

Pennsylvania
Pennsylvania Department of Health
Secretary of Health
Phone No. (717) 787–6436
Fax No. (717) 787–0191

Puerto Rico
Puerto Rico Department of Health
Secretary of Health
Phone No. (787) 274–7602
Fax No. (787) 250–6547

Rhode Island
Rhode Island Department of Health
Director of Health
Phone No. (401) 277–2231
Fax No. (401) 277–6548

South Carolina
SC Department of Health and Environmental Control
Commissioner
Phone No. (803) 734–4880
Fax No. (803) 734–4620

South Dakota
South Dakota State Department of Health
Secretary of Health
Phone No. (605) 773–3361
Fax No. (605) 773–5683

Tennessee
Tennessee Department of Health
State Health Officer
Phone No. (615) 741–3111
Fax No. (615) 741–2491

Texas
Department of Health
Commissioner of Health
Phone No. (512) 458–7375
Fax No. (512) 458–7477

Utah
Utah Department of Health
Director
Phone No. (801) 538–6111
Fax No. (801) 538–6306

Vermont
Vermont Department of Health
Commissioner
Phone No. (802) 863–7280
Fax No. (802) 865–7754

Virgin Islands
Virgin Islands Department of Health
Commissioner of Health
Phone No. (340) 774–0117
Fax No. (340) 777–4001

Virginia
Virginia Department of Health
State Health Commissioner
Phone No. (804) 786–3561
Fax No. (804) 786–4616

Washington
Washington State Department of Health
Acting Secretary of Health
Phone No. (360) 753–5871
Fax No. (360) 586–7424

West Virginia
Bureau for Public Health
WV Department of Health & Human Resources
Commissioner of Health
Phone No. (304) 558–2971
Fax No. (304) 558–1035

Wisconsin
Division of Health
Wisconsin Department of Health and Family Services
Administrator
Phone No. (608) 266–1511
Fax No. (608) 267–2832

Wyoming
Wyoming Department of Health
Director
Phone No. (307) 777–7656
Fax No. (307) 777–7439

Appendix C:
Patient Isolation Precautions

Reproduced with permission from USAMRIID's Medical Management of Biological Casualties Handbook **2004** [64].

Standard Precautions

- Wash hands after patient contact.
- Wear gloves when touching blood, body fluids, secretions, excretions and contaminated items.
- Wear a mask and eye protection, or a face shield during procedures likely to generate splashes or sprays of blood, body fluids, secretions or excretions.
- Handle used patient-care equipment and linen in a manner that prevents the transfer of microorganisms to people or equipment.

Use care when handling sharps and use a mouthpiece or other ventilation device as an alternative to mouth-to-mouth resuscitation when practical.

Standard precautions are employed in the care of ALL patients.

Airborne Precautions

Standard Precautions plus:

- Place the patient in a private room that has monitored negative air pressure, a minimum of six air changes/hour, and appropriate filtration of air before it is discharged from the room.
- Wear respiratory protection when entering the room.
- Limit movement and transport of the patient. Place a mask on the patient if they need to be moved.

Conventional Diseases requiring Airborne Precautions: Measles, Varicella, Pulmonary Tuberculosis.

Biothreat Diseases requiring Airborne Precautions: Smallpox.

Droplet Precautions

Standard Precautions plus:

- Place the patient in a private room or cohort them with someone with the same infection. If not feasible, maintain at least 3 feet between patients.
- Wear a mask when working within 3 feet of the patient.
- Limit movement and transport of the patient. Place a mask on the patient if they need to be moved.

Conventional Diseases requiring Droplet Precautions: Invasive Haemophilus influenzae and meningococcal disease, drug-resistant pneumococcal disease, diphtheria, pertussis, mycoplasma, GABHS, influenza, mumps, rubella, parvovirus.

Biothreat Diseases requiring Droplet Precautions: Pneumonic Plague.

Contact Precautions

Standard Precautions plus:

- Place the patient in a private room or cohort them with someone with the same infections if possible.

- Wear gloves when entering the room. Change gloves after contact with infective material.
- Wear a gown when entering the room if contact with patient is anticipated or if the patient has diarrhea, a colostomy or wound drainage not covered by a dressing.
- Limit the movement or transport of the patient from the room.
- Ensure that patient-care items, bedside equipment, and frequently touched surfaces receive daily cleaning.
- Dedicate use of noncritical patient-care equipment (such as stethoscopes) to a single patient, or cohort of patients with the same pathogen. If not feasible, adequate disinfection between patients is necessary.

Conventional Diseases requiring Contact Precautions: MRSA, VRE, Clostridium diffi-cile, RSV, parainfluenza, enteroviruses, enteric infections in the incontinent host, skin infections (SSSS, HSV, impetigo, lice, scabies), hemorrhagic conjunctivitis.

Biothreat Diseases requiring Contact Precautions: Viral Hemorrhagic Fevers.

For more information, see: Garner JS. Guideline for Infection Control Practices in Hospital Epidemiol 1996;17:53–80.

3
Care of Children in the Event of Bioterrorism

Subhash Chaudhary

3.1
Introduction

Bioterrorism is defined as the intentional use of microorganisms and/or their toxins to hurt others. For centuries, biological agents have been used throughout the world during wars [1]. The September 11, 2001, attacks on The World Trade Center and the October, 2001, mail delivery of letters or packages containing *Bacillus anthracis* spores brought the threat of chemical and bioterrorism to the civilian population to the forefront [2, 3]. Since then, much progress has been made in reducing vulnerability to such terrorist activity. Despite these measures, it is unlikely that future terrorist events will be prevented. In addition to maximum efforts at prevention, it is equally important that we learn to manage the consequences of bioterrorist activity and try to minimize adverse effects on the physical and mental well-being of our population, especially infants, young children, and adolescents.

3.2
Increased Vulnerability of Children

Children may sustain more serious harmful effects compared with adults after exposure to a biological agent. They are, therefore, clearly more vulnerable. Children may be exposed to these threats while at school, at home, or in other public areas [4]. During the terrorist bombing of the Alfred P. Murrah Federal Building in Oklahoma City, 10% of the victims were infants and children. Of these, 31% of the children died, compared with 21% of the adult victims [5]. During the industrial disaster that resulted in the release of methyl isocyanate in a heavily populated area in Bhopal, India, 20% of the patients admitted to a hospital on the first day were children younger than 15 years of age [6].

Bioterrorism Preparedness. Edited by Nancy Khardori
Copyright © 2006 WILEY-VCH Verlag GmbH & Co. KGaA, Weinheim
ISBN: 3-527-31235-8

Five major forms of terrorism have been identified by the American Academy of Pediatrics' Task Force on Terrorism, for which we need to be alert and prepare for as health care professionals, community leaders, and as individuals:

- thermomechanical
- chemical
- biological
- radiation
- psychological

Because children are physically smaller than adults, they can potentially get sicker from the same amount of exposure to a chemical or biological agent. Besides, their unique anatomic, physiological, and developmental features pose disproportionately higher risk from harmful substances [4, 7]. Some of these features are listed below.

3.2.1
Anatomic and Physiological Features Placing Children at Increased Risk of Vulnerability

1. **Increased Respiratory Rate:** The normal respiratory rate in infants and children is higher than in adults. For example, the respiratory rate of a 2-year-old child is essentially double that of an adult (30–32 min^{-1} compared with 16 min^{-1}) which increases the inhaled dose of vaporized agent to which they are exposed [8].

2. **Low Height:** Adult height is approximately twice that of a two-year-old child. The breathing zone for adults is usually 4 to 6 feet above ground whereas for children it is closer to the ground depending on the height and mobility of the child. Many of the highly toxic nerve agents are both volatile and heavy (sarin has a vapor density 4.86 times that of water) [7, 8]. As a result, they settle close to the ground directly in the breathing zone of young children.

3. **Thin, Permeable Skin:** The skin of young children is thinner than that of adults because of less keratinization. As a result, agents that can be absorbed through the skin are more likely to penetrate the thin skin of a child than the thick skin of an adult [9, 10].

4. **Large Surface-to-Mass Ratio:** In comparison with adults, children have a larger surface-to-mass ratio. A normal two-year-old child, weight 12 kg, height 85 cm, has a surface area of 0.55 m^2 compared with approximately 1.73 m^2 for a 70 kg adult. Although the child's weight is one-sixth that of an adult, the surface area is just one-third that of an adult. Thus, in addition to greater permeability, children have a large skin surface area through which chemicals and toxic agents can be absorbed [4, 11].

5. **Less Fluid Reserve:** Young children have less fluid reserve than adults, and are thus more prone to dehydration and shock. Agents that cause diarrhea and/or vomiting can, therefore, more easily lead to hypovolumic shock in young children [4, 11].

6. **Circulating Blood Volume:** Young children have less blood volume than do adults. As a result, loss of a relatively small amount of blood without intervention in a young child can more readily disturb the physiological balance and lead to progressive shock [11, 12].

7. **Skin Abrasions and Cuts:** Children, especially boys, because of their immaturity and rough playing habits, are more prone to have cuts and abrasions that can serve as portals of entry for toxins and microbes. Agents that penetrate the skin thus have yet another possible means of causing more harm to children than to adults.

3.2.2
Developmental Factors Involved in Increased Vulnerability of Children

1. **Limited Motor Skills:** Motor skills in infants, toddlers, and even young children are not fully developed. As a result they have difficulty escaping the scene of terrorist activity [4, 12].

2. **Immature Cognitive Ability:** Children who are able to ambulate may still not have the cognitive ability required to size up the situation and run away from the site of danger or terrorist activity. When exposed to noxious gases, young children may not understand the need to protect their faces and eyes [4, 12]. Moreover, because of their small size and height they are more likely to be trapped or left behind.

3. **Difficulty Following Directions:** Young children are more likely to have difficulty understanding and following directions regarding escape [12].

4. **Psychological Injury:** Children exposed to violence or biological–chemical terrorism, or those living under the threat of biological–chemical terrorism, are at risk of psychological injury [13, 14]. Children may witness parental injury and/or death during such events. This would result in both short and long-term psychological injury, including poor school performance [13, 15]. Because children think qualitatively differently from adults, they may not have the ability to understand what has happened. Studies have shown that children after observing or experiencing a violent event develop symptoms of post traumatic stress disorder (PTSD), though only a few of them develop clinically significant PTSD [16]. If a child loses a parent, there may be loss of emotional and financial support. With the loss of both parents, alternative placement may be necessary, resulting in a high risk of trauma because of relocation and adjustment to new caretakers [12]. In addition, the psychological and emotional health of a child depends largely on the emotional well being of the caretakers. If the parents maintain calmness and provide stability and a sense of security, the child is more likely to do well.

3.2.3
Delayed Diagnosis in Children

Rapid diagnosis in children may be hampered because the child is unable to describe his or her symptoms verbally [17]. In the event of bioterrorism, this problem may be intensified, especially if their parents were separated from the child or if they died in the attack.

3.2.4
Unique Management Needs of Children

Children, besides being more vulnerable, respond differently from adults. In children, because of smaller reserves of blood and fluid, their condition can worsen rapidly [12].

3.2.5
Decontamination Showers

Because children have larger surface area, they are prone to faster heat loss than adults. Cold water or outdoor decontamination showers may result in hypothermia in children. Therefore, warm water decontamination showers with heating lamps should be used to prevent hypothermia. Timing is critical in decontamination, however, so it should not be delayed unnecessarily to obtain warm water. Alternatively, children may be covered with heated blankets after showers [12, 18]. During such critical moments, parents will be hesitant to be separated from their children. During triage and decontamination allowing a parent to stay with the child may provide a dual benefit – this may eliminate the need for supervision and may aid thermoregulation for affected infants.

3.2.6
Doses of Medication

Children need individualized doses of medication depending on their weight and surface area. Some medications (e.g. chloramphenicol, tetracycline) also have different effects on developing children [4].

3.2.7
Size of Equipment

Because children are physically smaller than adults, they need equipment that has been specifically designed for their size (e.g., needles, tubing, oxygen masks,

ventilators, imaging and laboratory technology) [12]. It is imperative that ambu-
lances and emergency departments have appropriate pediatric equipment available.

3.2.8
Training of Healthcare Workers to Meet the Special Needs of Children

Health-care workers caring for children should be familiar with signs and symptoms indicative of a life-threatening condition so that timely intervention and
treatment can be provided. It is often difficult to perform some procedures (e.g.
intubation, insertion of intravenous line) on infants and young children. This
situation can be exacerbated in a bioterrorist event, because of the protective
equipment health-care workers may employ [12, 19]. Health-care workers involved
in emergency situations would therefore benefit from additional training in procedures specific to children. It is essential that pediatric healthcare workers and
institutions that take care of children are involved in planning such activities at
every level [4].

3.2.9
Communication with Children about Disasters

It is extremely important for parents and care-givers to convey the message to
children that they are safe. Children can feel frightened and scared by the scenes
shown on television. Parents should not dismiss child's fears. They should explore
the issues and provide facts. So it is very important to let them know that these are
isolated events in certain areas and they will not be hurt. Children need repeated
reassurances that everything has been done to keep them safe. Parents must talk to
their children about the events at a level the children can understand. They need to
know that terrorists and acts of terrorism are bad but all the people of that faith,
religion, or culture are not bad. Parents should make sure their children view such
acts of horror or terrorism on television in their company so that proper explan-
ation can be given to them. Care-givers also need to make sure children are not
watching such events on television repeatedly, because media have a tendency to
repeat such events to the point of exhaustion [20].

3.2.10
Communication with Adolescents about Disasters

Because such events can have major adverse effects on adolescents, it may be even
more important to communicate with them very regularly in the event of disaster
or terrorist activity. They may not verbalize fears but can develop sleep disturban-
ces, tiredness, and lack of energy and interest in activities they used to enjoy. They

may start using illicit drugs. Others, especially boys, may vent their anger toward members of a particular religious or ethnic group [20, 21].

3.3
Categories of Biological Agents and Toxins

Clinical manifestations of diseases caused by agents of bioterrorism in children may be in the form of an acute rash with fever, acute respiratory distress and fever, influenza-like illness, neurological syndromes, or blistering syndromes [17].

According to the Centers for Disease Control and Prevention there are currently approximately thirty-five potential bioweapons. The hidden nature of the release of a bioweapon, as was apparent for the anthrax-containing letters in September and October 2001, has much potential to cause large scale fear and anxiety. These biological agents have been divided into three categories (A, B, and C), depending on the potential terrorist threat they may pose. Category A includes agents that are either extremely contagious or easy to disseminate, have very high mortality rate, and require highest priority public-health response and preparedness [22]. Smallpox, anthrax, plague, botulism, tularemia, and viral hemorrhagic fevers belong to this category. Most of the Category A agents will be discussed in this chapter with reference to children. The goal is to emphasize differences, where they exist, between the clinical picture, and management, for children and adults.

3.3.1
Smallpox (Variola)

Smallpox, once prevalent worldwide, was eradicated in 1977 by a global campaign under the leadership of the World Health Organization (WHO). In 1980 the World Health Assembly, the supreme governing body of the WHO, recommended that all the countries should stop smallpox vaccination [23]. As a result, the vaccine was removed from the commercial market in 1983 and production was discontinued. Terrorist events over recent years have raised serious concerns about smallpox as a possible bioterrorism agent [24].

Smallpox virus is one of the largest and most complex DNA viruses. Infection occurs after the virus infects the oropharyngeal or respiratory mucosa of a susceptible child. After an incubation period of approximately 2 weeks (range 7–17 days) children commonly present with high fever, malaise, prostration, backache, and headache. Rash begins as maculopapular lesions in the month and pharynx, face, and forearms. It then spreads to the trunk and legs. Within 1 to 2 days the rash becomes vesicular and then pustular. Lesions are round, tense, and deep. Crusts which eventually scab are seen 8 to 9 days after the onset of the rash. A mortality rate of up to 30% is reported and spread from person to person is easy through respiratory droplets, aerosols, and the scabs [24, 25].

In the United States, vaccination was recommended for all children at age 1 to 2 years and in endemic areas was given to infants as young as 3 months. Routine vaccination was discontinued in the US in 1972. As a result, all children and adults younger than 33 years of age are susceptible [26].

3.3.1.1 Bioterrorism Potential

The history of smallpox, nature of the virus, and its known mode of transmission make smallpox a very likely agent of bioterrorism [24]. There are reports that in the former Soviet Union smallpox was being developed for biological warfare. There are also concerns that the virus, the expertise, and the equipment might have fallen into non-Russian hands [27]. It is well established that smallpox virus can be produced in large quantities and is stable for storage and transportation. It can also be released as a stable aerosol. All these features make it a suitable bioweapon. Currently, when all children and most adults have little or no immunity, a single case of smallpox will be a public health and medical emergency [26].

3.3.1.2 Differentiation from Varicella Rash

Many physicians have never seen a case of smallpox but they have seen children with chickenpox (Varicella). There is potential for misdiagnosis of initial cases of bioterrorism-associated smallpox as being chickenpox. Varicella rash is characterized by superficial lesions in multiple crops and in various stages of development (macules, papules, vesicles, and pustules) mainly on scalp, face, and trunk. In general, the child does not look sick or toxic. Most lesions change to crust forms in 6–8 days and crusts fall off within 14 days of the onset of rash [26].

3.3.1.3 Antiviral Therapy

There is currently no effective antiviral therapy for smallpox. Cidofovir, a nucleoside analog DNA polymerase inhibitor, is licensed for the treatment of cytomegalovirus retinitis. There are limited in-vitro and animal data which show it has some activity against the smallpox virus, but data to support its efficacy in treating smallpox are not available. Vaccinia immune globulin has no role in the therapy of smallpox [28].

3.3.1.4 Infection Control and Isolation of Hospitalized Child

Any child suspected of having smallpox should be maintained with standard, contact, and airborne precautions. The child should be in a private room with negative pressure ventilation and high-efficiency particulate air filtration. Health-care workers, even if successfully immunized, must wear an N95 or higher-quality respirator, gloves, and gown before entering the child's room. To reduce the chances of transmission by fomites, the patient should wear a mask and be covered with sheets when leaving the room. Laundry and waste should be placed in biohazard bags and autoclaved. The patient's bedding and clothing should either be incinerated or washed in hot water with laundry detergent followed by hot-air drying. Terminal decontamination of the room using sodium hypochlorite or quaternary ammonia compounds is necessary. Infection control personnel should be involved from the very beginning [28].

3.3.1.5 Smallpox Vaccine

The vaccine is a lyophilized live vaccinia virus preparation that is given by using a bifurcated needle to deliver vaccine into epidermis. It causes a local infection and is accompanied by pruritus. Patient develops fever, malaise, and regional lymphadenitis approximately one week after vaccination. Occasionally a child (more likely immunosuppressed) can develop life-threatening progressive lesions or disseminated infection. In those with chronic skin conditions it can lead to severe eczema vaccinatum. Vaccinia immune globulin can reduce the severity of many of these complications [24].

3.3.1.6 Strategies for Vaccination in the Event of Bioterrorism

If smallpox is used as a bioweapon and released as an aerosol it has the potential to disseminate very quickly and widely and lead to several first-generation cases of smallpox followed by severalfold second-generation cases [24]. Three major strategies have been considered:

1. mass pre-event vaccination
2. voluntary vaccination
3. ring vaccination

The first two strategies have the disadvantage of exposing millions of people to serious possible risks of vaccinia immunization. It is well established that vaccination within 3 to 4 days of exposure gives some protection against smallpox and significant protection against fatal disease. In a ring vaccination strategy there is no pre-event vaccination. When a smallpox case is diagnosed the patient is isolated and vaccination is given to all those who had face-to-face household contact; those who have been in close proximity to a person with active smallpox skin lesions, or those who have been in contact with such an exposed person. Vaccination must occur within 3 to 4 days of exposure. Advantages of this strategy are that it exposes the minimum number of persons to adverse reactions of vaccination and enables best use of limited supplies of the vaccine. It requires the help of very effective, well trained, and immediately available public health system. Ring immunization strategy was used successfully in developing countries during the smallpox-eradication program [24, 26]. The American Academy of Pediatrics supports the ring vaccination strategy in cases of bioterrorism [29].

3.3.2
Anthrax

Anthrax, an infectious disease caused by *Bacillus anthracis*, has become a rare disease in the developed world. Children, like adults, can develop cutaneous, inhalation, or gastrointestinal manifestations; the cutaneous type is the most common [30].

Cutaneous manifestations of anthrax occur when spores (viable in the soil for years) enter at sites of broken skin. After an incubation period of 1 to 7 days a painless papule appears which is complicated by the appearance of vesicles containing clear or serosanguinous fluid around it over the next few days. Subsequent ulceration in 1 to 2 days leads to formation of a central black eschar. Discharge from skin lesions is potentially infectious but person-to-person transmission in children has occurred rarely [17, 30].

Inhalational anthrax is the rapidly progressive form of the disease. Incubation period is usually 1 to 6 days but can be as long as 60 days. During the prodromal phase the child has fever, chills, non-productive cough, chest pain, headache, and muscle aches. Two to five days later the patient develops hemorrhagic pleural effusion, bacteremia, and toxemia which can lead to severe respiratory difficulty, hypoxia, and shock [30].

Gastrointestinal anthrax may have oropharyngeal or intestinal manifestations. The oropharyngeal form may present with posterior oropharyngeal ulcers which typically are unilateral and are associated with prominent neck swelling, enlarged nodes and septicemia. The intestinal variety may start with upper gastrointestinal symptoms, for example anorexia, nausea, vomiting, and fever, and then progress to severe abdominal pain, massive ascitis, and blood in vomitus and stools. The mortality rate in inhalational and gastrointestinal anthrax is more than 50% but for cutaneous form with appropriate treatment is less than 1% [30, 31].

3.3.2.1   Bioterrorism Potential
Spores of *Bacillus anthracis* are highly stable and can infect via the respiratory route, and inhalational anthrax has a very high mortality. For these reasons *Bacillus anthracis* has high potential for use as a bioweapon. Besides aerosolization, these spores can be introduced into food products or water supplies and pose health risks. In 2001, after intentional contamination of the mail, twenty-two cases of anthrax (eleven inhalational, eleven cutaneous) occurred in the United States. Of these twenty-two patients, there was one pediatric patient, a seven-month-old infant who developed cutaneous anthrax associated with systemic complications [31].

3.3.2.2   Case Report of Infant with Bioterrorism-associated Cutaneous Anthrax
This, apparently healthy infant with no significant medical history acquired anthrax from his mother's workplace during a stay of one hour. The skin lesion

started on his left upper extremity as a red macule which progressed to a papular lesion with marked edema despite antibiotic therapy (Fig. 3.1) He was admitted to the hospital and given intravenous antibiotic therapy and had incision and drainage. He developed the systemic complications hyponatremia, coagulopathy, and severe microangiopathic hemolytic anemia. These complications are uncommon with cutaneous anthrax but are well described with envenomations, especially that of *Loxoscales recluse*. Spores of *Bacillus anthracis* at his mother's workplace most probably led to his cutaneous infection. Spores on hands of some workers probably entered his skin via inapparent skin abrasion on an exposed area [32].

The initial lesion of cutaneous anthrax may be confused with cellulites or insect bite. Differentiating features in cutaneous anthrax are the prominent edema and lack of pain. As seen in this case, cutaneous anthrax in infants and young children can rapidly progress to severe disease with systemic complications. For this reason infants and young children with even presumptive diagnosis of cutaneous anthrax should be admitted to the hospital. After appropriate cultures and investigations, they should receive intravenous antibiotic therapy and be monitored closely for electrolyte and hematologic complications [32].

3.3.2.3 Treatment

Inhalational anthrax (associated with bioterrorism)

Historically penicillin was the treatment of choice. There are data showing that *Bacillus anthracis* strains produce a cephalosporinase and that some strains contain an inducible beta-lactamase. So presumptive treatment for inhalational anthrax in children should be ciprofloxacin $10-15$ mg kg^{-1} every twelve hours (maximum 1 g day^{-1}) intravenously or doxycycline 2.2 mg kg^{-1} every twelve hours (maximum 200 mg day^{-1}). If the clinical picture is suggestive of meningitis, ciprofloxacin should be used instead of doxycycline because of better penetration of ciprofloxacin into the central nervous system. One or two of the following antibiotics rifampin, vancomycin, penicillin, ampicillin, chloramphenicol, imipenem, clindamycin, and clarithromycin, which have in-vitro activity, should also be given with ciprofloxacin or doxycycline for initial therapy. Clindamycin (30 mg kg^{-1} day^{-1} intravenously) may be preferable to the others, because its inhibition of protein synthesis may inhibit production of anthrax toxins (lethal factor, edema factor, and protective antigen). When the patient is stable and improving therapy can be changed to oral ciprofloxacin or doxycycline in same doses to complete 14–21 days of therapy. At that stage, because of concern about potential adverse effects of continued use of ciprofloxacin or doxycycline, therapy may be changed to oral amoxicillin 80 mg kg^{-1} day^{-1} divided into doses every eight hours (not to exceed 1500 mg day^{-1}) to complete 60 days of therapy if the isolate involved is susceptible to penicillin [31, 33, 34].

Cutaneous anthrax (associated with bioterrorism)

Children with cutaneous lesion and signs of systemic involvement, marked edema, or with lesions involving head and neck, and those under the age of 2 years should

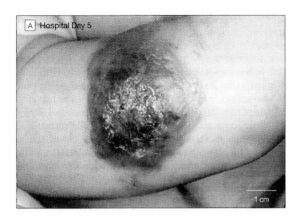

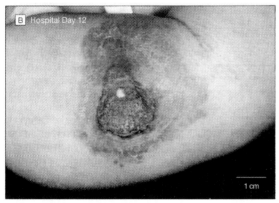

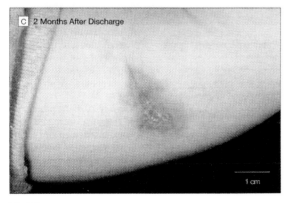

Fig. 3.1
Lesion of bioterrorism-associated cutaneous anthrax. (Reproduced with permission from Ref. [32].)

be managed initially with intravenous therapy and a multidrug regimen as described for inhalational disease. Steroids also may be used for these children. Although 7 to 10 days of therapy may be adequate for the disease, therapy is recommended for 60 days, because there is risk of simultaneous aerosol exposure. After 7 to 10 days antibiotic therapy may be changed to oral amoxicillin 80 mg kg^{-1} day^{-1} divided into doses every eight hours (maximum 1500 mg day^{-1}) to complete 60 days of therapy if the isolate involved is susceptible to penicillin [31, 33, 34].

Post-exposure prophylaxis in children

Ciprofloxacin or doxycycline orally in the same dosage schedules as for treatment for 60 days are optimum. Amoxicillin may be given in same dosage schedule as for treatment for 60 days if the isolate involved is susceptible to penicillin [31, 33, 34].

Infection control and isolation precautions

Standard precautions should be followed. Contaminated dressings and bedclothes should be incinerated or steam-sterilized to destroy spores. The patient's room must also be disinfected to kill spores. Special precautions to prevent transmission must be taken when performing autopsies on patients who died from systemic disease [33].

Vaccine

A cell-free culture filtrate-absorbed vaccine is available for persons at risk of repeated exposure to *Bacillus anthracis* spores. This vaccine is not licensed for post-exposure use. This vaccine is not licensed for use with children. No data are available on the effectiveness or safety of this vaccine in children [33].

3.3.3
Botulism

Botulism is a neuroparalytic disorder caused by one of the seven antigenic toxin types (A through G) of *Clostridium botulinum*. Most human botulism cases are caused by neurotoxins A, B, and E but, rarely, F neurotoxin can also cause disease [35]. The four known types of human botulism are discussed below.

3.3.3.1   Foodborne Botulism

This variety of illness results when food contaminated with spores of *C. botulinum* is preserved or stored improperly under anaerobic conditions. This leads to germination, multiplication, and toxin production. Ingestion of preformed toxin leads to illness. In the US most foodborne outbreaks of botulism have been attributed to toxin types A, B, and E. In this form of illness even after severe disease there is no immunity to toxin [35].

3.3.3.2   Wound Botulism

This type of illness results when *C. botulinum* bacteria contaminate traumatized tissue. After colonization, bacteria multiply and produce toxin. In addition to classical crush injury, in recent years injection of contaminated black tar heroin has been associated with many reported cases [35].

3.3.3.3   Infant Botulism

Infant botulism is the most recently recognized (since 1976) and the most common type of human botulism in the United States. This form results after ingested spores of *C. botulinum* and related species germinate, colonize the colon of young infants, and produce neurotoxin most probably as a result of transient permissiveness of the intestinal microflora of the infant. Typically it is seen in infants younger than 6 months of age, has relatively longer incubation period (3 to 30 days) and relatively slow onset of symptoms. Constipation often precedes reduced activity, loss of facial expression, slow feeding, weak cry, poor gag reflex, generalized weakness, hypotonia, and apnea [35, 36].

3.3.3.4   Botulism of Undetermined Etiology

This rare variety is typically seen in patients older than 1 year. Clinical manifestations are similar but no wound or food source can be identified [36].

3.3.3.5   Bioterrorism Potential

Botulinum toxin is one of the CDC Category A agents of bioterrorism. In natural disease airborne spread of preformed toxin does not occur. For bioterrorist activity aerosolization of the toxin will be the most probable route, although sabotage of the food supply is a possibility. Irreversible binding of toxin to peripheral cholinergic synapses results in block of release of acetylcholine and paralysis. In the United States, natural disease is seen as small clusters or single cases. Clinical features of the disease acquired by airborne transmission should be same as for foodborne disease. Blurred vision, diplopia, and dry month can be the early features followed by descending flaccid paralysis. A high index of suspicion is critical for early diagnosis. Features which should be clues to the possibility of bioterrorist activity include [38]:

1. simultaneous occurrence of several cases in children and adults; and
2. outbreaks due to unusual botulinum toxin types (C, D, F, G)

3.3.3.6   Supportive Care

Meticulous respiratory and nutritional care are the mainstay of treatment of all varieties of botulism.

3.3.3.7 Antitoxin

Passive antibody given on early suspicion of botulism can minimize the chance of subsequent nerve damage, and severe disease. Antitoxin does not reverse existing paralysis. Human-derived botulinum antitoxin (botulism immune globulin intravenous (human) (BIG-IV)) has been shown to be beneficial in reducing the number of days of hospitalization, mechanical ventilation, and tube feeding for those suffering from infant botulism [37]. Trivalent (types A, B, and E) and bivalent (types A and B) equine botulism antitoxins are available from the Centers for Disease Control and Prevention through state health departments for management of all forms of botulism. With equine antitoxin there is a risk of immediate hypersensitivity and sensitization to equine proteins. These preparations have been used in children without apparent short-term adverse reactions. The United States Army has a heptavalent antitoxin preparation under investigation which could potentially be used in a bioterrorist attack with one of the unusual types of toxin [37].

3.3.3.8 Antibiotics

Antimicrobial agents have no effect on the toxin and should be avoided in infant botulism because lysis of *C. botulinum* may potentially increase the level of the toxin. Aminoglycoside antibiotics and clindamycin should especially be avoided because they can worsen neuromuscular blockade [36, 37].

3.3.3.9 Infection Control and Isolation of Hospitalized Children

Standard precautions should be followed. No special precautions are needed.

3.3.3.10 Decontamination

Botulinum toxin is easily destroyed by extremes of temperature and humidity. Two days after aerosolization, significant inactivation of toxin occurs. The toxin does not penetrate intact skin. Skin and clothing should be washed with soap and water after exposure to toxin. Bleach solution (0.1%) may be used to clean contaminated objects and surfaces [36].

3.3.3.11 Care of Exposed Persons

Equine antitoxin is not currently recommended for post-exposure prophylaxis.

3.3.3.12 Vaccine

Investigational pentavalent (A, B, C, D, E) botulinum toxoid vaccine has been used for laboratory workers at high risk of exposure and for military personnel. Vaccine has not been studied in children [37].

3.3.4
Plague

Yersinia pestis, a pleomorphic, bipolar staining, Gram-negative coccobacillus is the causative agent of plague. Plague is a zoonotic infection of rodents, carnivores, and their fleas. In nature, most common manifestation in children is the bubonic variety with acute onset of fever and painful swollen lymph nodes (bubos), most commonly in the inguinal area. It is generally transmitted by infected fleas, through their bites, and occasionally by direct contact with the fluids or tissues of infected rodents or mammals, including domestic cats. The incubation period for bubonic plague is 2 to 6 days. Less common manifestations include a septicemic form (acute respiratory distress, hypotension, and disseminated intravascular coagnlopathy), a pneumonic form (fever, cough, hemoptysis, and dyspnea), and meningitis [38].

The septicemic form manifests as acute respiratory distress, hypotension, and disseminated intravascular coagulation. This variety is seen mostly as a complication of bubonic plague but can also occur as a result of direct contact with infectious materials or the bite of an infected flea. Pneumonic form is acquired by inhalation of respiratory droplets from an animal or human with respiratory plague or from aerosol exposure in the laboratory. The incubation period for this type is 2 to 4 days. Meningitis is seen rarely. High fever, chills, headache, and rapidly advancing weakness are common features of all forms of plague [38, 39].

3.3.4.1   **Bioterrorism Potential**
Because of its high contagiousness, plague is regarded as a major potential bioterrorism threat. In this setting most common manifestations will be the pneumonic variety with or without the septicemic form. Presentation of many previously healthy children and adults with rapidly progressive pneumonia and hemoptysis should raise clinical suspicion of plague of bioterrorist origin [17, 40].

3.3.4.2   **Infection Control and Isolation Precautions**
Patients should be placed in isolation with droplet precautions until pneumonia has been ruled out and appropriate antibiotic therapy started. In patients with pneumonic plague droplet precautions should be maintained for 48 h after starting appropriate antibiotic treatment. Microbiology laboratories should be notified before sending specimens from a patient with suspect plague, so that proper precautions can be taken to minimize the risk to laboratory personnel [39, 40].

3.3.4.3   **Treatment**
For acutely ill presumed plague patients and in the event of bioterrorism in a contained casualty setting streptomycin (15 mg kg^{-1} IM twice daily with maximum

daily dose of 2 g day^{-1}) and gentamicin (2.5 mg kg^{-1} IM or IV every 8 h) are the drugs of choice. Gentamicin 5 mg kg^{-1} IM or IV once daily may also be used. Alternative choices are doxycycline (2.2 mg kg^{-1} IV twice daily with maximum 200 mg day^{-1}), ciprofloxacin (15 mg kg^{-1} IV twice daily with maximum 1 g day^{-1}), and chloramphenicol (25 mg kg^{-1} IV every 6 h). Plasmid-mediated multiple antibiotic resistance in Y. pestis has been known. In-vitro antibiotic susceptibility should be determined and antibiotic therapy modified accordingly. Chloramphenicol is the antibiotic of choice for patients with meningitis. Duration of therapy should be at least seven days but in complicated cases it may be longer. In pneumonic plague there is high morbidity and mortality.

In a mass casualty setting oral therapy with doxycycline (2.2 mg kg^{-1} orally twice daily; maximum 200 mg day^{-1}) or ciprofloxacin (20 mg kg^{-1} twice daily; maximum 1 g day^{-1}) may be used. Oral chloramphenicol (25 mg kg^{-1}, orally, four times daily) is another alternative in this setting [39, 40].

3.3.4.4 Post-exposure Prophylaxis

If there is an outbreak of plague in a community, parenteral antibiotics should be started for all patients with a temperature of 38.5 °C or higher or new cough and for infants with tachypnea. When it is not possible to give parenteral therapy, oral antibiotics as recommended for treatment in a mass casualty setting should be used [39, 40].

Asymptomatic patients in close contact (household, hospital) with a patient with pneumonic plague should be given antibiotic prophylaxis for seven days with doxycycline or ciprofloxacin at the same doses as for treatment in mass casualty settings. Chloramphenicol is an alternative choice for this purpose [39, 40].

3.3.5
Tularemia

Francisella tularensis, the causative agent of tularemia, is a nonmotile, aerobic, Gram-negative pleomorphic coccobacillus. Wild animals (e.g. rabbits, muskrats, and moles), domestic animals (e.g. sheep, cattle, and cats), arthropods that bite these animals (e.g. deerflies, mosquitoes, and ticks), and contaminated soil and water can serve as the sources of this pathogen. Ticks and infected animals are contagious for prolonged periods. Ticks and rabbits are the major source of infection in humans in the United States. As a result, most cases are seen in summer months.

Most children present with acute onset illness in the form of fever, chills, headache, and muscle aches after an incubation period of 3 to 5 days. The two most common syndromes are the ulceroglandular form and the glandular variety. Less common manifestations are the occuloglandular, oropharyngeal, typhoidal, intestinal, and pneumonic types. Sepsis may complicate any of these syndromes [41]. Pulse–temperature dissociation (inadequate rise in pulse rate in proportion to rise in body temperature) is seen in several patients [42].

3.3.5.1 Bioterrorism Potential

F. tularensis is a CDC category A potential bioterrorism agent because it is highly infectious (inhalation or inoculation of ten or more organisms can cause disease), is easy to disseminate, and can lead to serious illness and death. Inhalation of aerosols will be the most likely route of infection if this organism is used for bioterrorist activity. There is no human-to-human spread. In a bioterrorist attack it will probably be released in a heavily populated region (urban area). Initial manifestations may resemble that of influenza or atypical pneumonia. Rapid progression from upper respiratory symptoms to life-threatening lower respiratory disease with systemic manifestations should raise suspicion of a possible bioterrorism agent such as *F. tularensis*. Inhalational tularemia as a natural disease is seen mostly in rural populations. Diagnosis of inhalation disease in urban areas and in children without skin or lymph node manifestations should raise concern about bioterrorist activity. In general, progression of the disease is slower and mortality rates are lower in tularemia as compared with those of inhalational anthrax and plague [17, 43].

3.3.5.2 Post-exposure Prophylaxis

Because there is no human-to-human spread, close contacts of a patient do not need prophylaxis.

In the event of bioterrorism, if early in the incubation period it is known that the bioweapon is *F. tularensis* and exposed persons can be identified and reached, oral doxycycline or ciprofloxacin may be given in therapeutic doses for 14 days. If bioterrorism-associated tularemia patients have been identified, exposed children with unexplained fever or upper respiratory illness within 14 days of presumed exposure should be given antibiotic therapy as if they had tularemia [43].

3.3.5.3 Isolation Precautions and Infection Control

Standard precautions should be followed when a child with *F. tularensis* infection is in a healthcare facility. When presumptive diagnosis of tularemia is made in a child, microbiology laboratory personnel should be informed so that proper precautions (face mask, rubber gloves, and biological safety cabinet) are followed when handling clinical specimens or cultures. Body surfaces and clothes should be washed with soap and water if there has been direct exposure to powder or liquid aerosols containing *F. tularensis*. During autopsy sawing of bones should be avoided because of potential aerosol spread [44].

In cases of laboratory spill or intentional use in which there is concern about environmental contamination the suspected contaminant should be sprayed with

10% bleach solution. After 10 min a 70% solution of alcohol may be used to clean the area further.

3.3.5.4 Treatment

Streptomycin, gentamicin, tetracycline, ciprofloxacin, and chloramphenicol are effective therapeutic agents against *F. tularensis*. Streptomycin (15 mg kg^{-1} IM, twice daily for 10 days; maximum 2 g day^{-1}) and gentamicin (2.5 mg kg^{-1} IM or IV every 8 h for 10 days) are the preferred antibiotics in a contained casualty situation [41, 43, 44]. Enderlin et al. [45] in their review concluded that use of streptomycin resulted in a better cure rate and very low relapse rate compared with gentamicin. Doxycycline (2.2 mg kg^{-1} IV twice daily for 10 days; maximum 200 gm day^{-1}), ciprofloxacin (15 mg kg^{-1} IV twice daily for 10 days; maximum 1 g day^{-1}), and chloramphenicol (15 mg kg^{-1} IV every 6 h for 14–21 days) may be used as alternatives. These antibiotics may be switched to oral route when the clinical condition of the child is better [41, 43, 44].

3.3.5.5 Vaccine

A live attenuated vaccine is under investigation and is available for the protection of laboratory workers. This vaccine has not been tested on children [44].

References

1 Christopher G, Cieslak T, Pavlin J, Eitzen E. **1997** Biological warfare: a historical perspective. *JAMA* 278, 412–417

2 Bush LM, Abrams BH, Beall A, Johnson CC. **2001** Index case of fatal inhalational anthrax due to bioterrorism in the United States. *N. Engl. J. Med*, 345:1607–1610

3 Jernigan JA, Stephens DS, Ashford DA, et al. **2001** Bioterrorism-related inhalation anthrax: The first 10 cases reported in the United States. *Emerg. Infect. Dis*, 7:933–944

4 American Academy of Pediatrics, Committee on Environmental Health and Committee on Infectious Diseases. **2000** Chemical–biological terrorism and its impact on children: a subject review. *Pediatrics*, 105:662–670

5 Mallonee S, Shariat S, Stennies G, Waxweiler R, Hogan D, Jordan F. **1996** Physical injuries and fatalities resulting from the Oklahoma City bombing. *JAMA*, 276:382–387

6 Mehta PS, Mehta AS, Mehta SJ, Makhijani AB. **1990** Bhopal tragedy's health effects. A review of methyl isocyanate toxicity. *JAMA*, 264:2781–2787

7 Wright JL. **2002** Testimony before the Senate Committee on Health, Education, Labor and Pensions Subcommittee on Children and Families [Internet]. Elk Grove Village, IL: American Academy of Pediatrics; c2002. Available from: http://www.aap.org/advocacy/washing/dr_wright.htm

8 Bearer C. **1995** How are children different from adults? *Environ Health Perspect*, 103:7–12

9 Slater MS, Trunkey DD. **1997** Terrorism in America: an evolving threat. *Arch. Surg*, 132:1059–1066

10 Momeni A, Aminjavaheri M. **1994** Skin manifestations of mustard gas in a group of 14 children and teenagers: a clinical study. *Int J Dermatol*, 33:184–187

11 Bernardo LM. **2001** Pediatric Implications in Bioterrorism, Part 1: Physiologic and Psychologic Differences. *Int. J. Trauma Nurs*, 7:14–16

12 The Youngest Victims: Disaster Preparedness to Meet Children's Needs [Internet]. Elk Grove Village, IL: American Academy of Pediatrics; c2002. Available from: http://www.aap.org/advocacy/releases/ disaster_preparedness.htm

13 Holloway HC, Norwood AE, Fullerton CS, Engel CC, Ursano RJ. **1997** The threat of biological weapons: prophylaxis and mitigation of psychological and social consequences. *JAMA*. 278:425–427

14 Hyams KC, Wignall FS, Roswell R. **1996** War syndromes and their evaluation: from the US Civil War to the Persian Gulf War. Ann Intern Med. 125:398–405

15 Hurt H, Malmud E, Brodsky NL, Giannetta J. **2001** Exposure to Violence: psychological and academic correlates in child witnesses. *Arch. Pediatr. Adolesc. Med.* 155:1351–1356

16 Berman SL, Kurtines WM, Silverman WK, Serafini LT. **1996** The impact of exposure to crime and violence on urban youth. Am J. Orthopsychiatry 66:329–331

17 Patt HA, Feigin RD. **2002** Diagnosis and Management of Suspected Cases of Bioterrorism: A Pediatric Perspective. *Pediatrics* 109:685–692

18 Rotenberg JS, Burklow TR, Selanikio JS. **2003** Weapons of Mass Destruction: The Decontamination of Children. *Pediatric Annals* 32, 4:260–267

19 Tucker JB. **1997** National Health and Medical Services response to incidents of chemical and biological terrorism. *JAMA*. 278:362–368

20 AAP Offers Advice on Communicating with Children about Disasters [Internet]. Elk Grove Village, IL: American Academy of Pediatrics; c2002. [cited 2005 Aug. 11]. Available from: http://www.aap.org/advocacy/releases/disastercomm.htm

21 Spitalny K, Gurian A, Goodman RF. **2002** Helping Children and Teens Cope with Traumatic Events and Death: The Role of School Health Professionals [Internet]. New York, NY: NYU Child Study Center. Available from: http://www.aboutourkids.org/aboutour/articles/crisis_school-health.html

22 Biological and chemical terrorism: strategic plan for preparedness and response. **2000** Recommendations of the CDC Strategic Planning Workgroup. *MMWR Morb Mortal Wkly Rep*, 49 (RR-4):1–14

23 World Health Organization. **1980** The Global Eradication of Smallpox: Final Report of the Global Commission for the Certification of Smallpox Eradication. Geneva, Switzerland: World Health Organization

24 Henderson DA, Inglesby TV, Bartlett JG, et al., for the Working Group on Civilian Biodefense. **1999** Smallpox as a biological weapon: medical and public health management. *JAMA*. 281:2127–2137

25 Cherry JD. **2004** Smallpox (Variola Virus). In Feigin RD, Cherry JD, Demmler GJ, Kaplan SL. (eds) Text Book of Pediatric Infectious Diseases. 5th edn. Philadelphia, Saunders, 1972–1977

26 Baltimore RS and McMillan JA. **2002** Smallpox and the Smallpox Vaccine Controversy. *Pediatr Infect Dis J*, 21:787–790.

27 Alibek K. **1999** Biohazard. New York, NY: Random House Inc

28 American Academy of Pediatrics. **2003** Smallpox (Variola). In Pickering LK, ed. Red Book: Report of Committee on Infectious Diseases. 26th edn. Elk Grove Village, IL: *American Academy of Pediatrics*, 554–558

29 American Academy of Pediatrics: Policy Statement, Committee on Infectious Diseases. **2002** Smallpox Vaccine. *Pediatrics*, 110:841–845

30 Edwards ME. **2004** Anthrax. In Feigin RD, Cherry JD, Demler GJ, Kaplan SL. (eds). Text Book of Pediatric Infectious Diseases. 5th edn. Philadelphia, Saunders, 1314–1318

31 Ingelsby TV, O'Toole T, Henderson DA, et al. **2002** Anthrax as a Biological Weapon. Updated Recommendations for Management: *JAMA*, 287:2236–2252

32 Freedman A, Afonja O, Chang M, et al. **2002** Cutaneous anthrax associated with microangiopathic hemolytic anemia and coagulopathy in a 7-month-old infant. *JAMA*, 287:869–874

33 American Academy of Pediatrics. **2003** Anthrax. In Pickering LK, ed. Red Book: Report of Committee on Infectious Diseases. 26th edn. Elk Grove Village, IL: *American Academy of Pediatrics*, 196–199

34 Update: Interim Recommendations for Antimicrobial Prophylaxis for Children and Breastfeeding Mothers and Treatment of Children with Anthrax. **2001** *MMWR Morb Mortal Wkly Rep.* 50:1014–1017

35 Arnon SS. **2004** Infant Botulism. In Feigin RD, Cherry JD, Demler GJ, Kaplan SL. (eds). Text Book of Pediatric Infectious Diseases. 5th edn. Philadelphia, Saunders, 1758–1766

36 American Academy of Pediatrics. **2003** Clostridial Infections: Botulism and Infant Botulism. In: Pickering LK, ed. Red Book: 2003 Report of Committee on Infectious Diseases. 26th edn. Elk Grove Village, IL: *American Academy of Pediatrics*, 243–246

37 Arnon SS, Schechter R, Inglesby TV, et al., for the Working Group on Civilian Biodefense. **2001** Botulinum Toxin as a Biological Weapon: Medical and Public Health Management. *JAMA*, 285:1059–1070

38 Goldstein MD. **2004** Plague. In Feigin RD, Cherry JD, Demler GJ, Kaplan SL. (eds). Text Book of Pediatric Infectious Diseases. 5th edn. Philadelphia, Saunders, 1487–1492

39 American Academy of Pediatrics. **2003** Plague. In Pickering LK, ed. Red Book: Report of Committee on Infectious Diseases. 26th edn. Elk Grove Village, IL: *American Academy of Pediatrics*, 2003:487–489

40 Inglesby TV, Henderson DA, Bartlet JG, et al., for the Working Group on Civilian Biodefense. **2000** Plague as a biological weapon: medical and public health management. JAMA, 283:2281–2290

41 Feigin RD and Lau CC. **2004** Tularemia. In Feigin RD, Cherry JD, Demler GJ, Kaplan SL. (eds). Text Book of Pediatric Infectious Diseases. 5th edn. Philadelphia, Saunders, 1628–1635

42 Evans ME, Gregory DW, Schaffner W, McGee ZA. **1985** Tularemia: a 30-year experience with 88 cases. Medicine, 64:251–269

43 Dennis DT, Inglesby TV, Henderson DA, et al., for the Working Group on Civilian Biodefense. **2000** Tularemia as a biological weapon: medical and public health management. *JAMA*, 283:2763–2773

44 American Academy of Pediatrics. **2003** Tularemia. In Pickering LK, ed. Red Book: 2003 Report of Committee on Infectious Diseases. 26th edn. Elk Grove Village, IL: *American Academy of Pediatrics*, 666–667

45 Enderlin G, Morales L, Jacobs RF, et al. **1994** Streptomycin and alternative agents for the treatment of tularemia: Review of the literature. *Clin. Infect. Dis.* 19:42–47

4
Smallpox: Virology, Clinical Presentation, and Prevention

James M. Goodrich

4.1
Introduction

World events have increased interest in biological agents as possible terrorist weapons. Smallpox has been mentioned as a possible agent for use in a terrorist attack and as a terror weapon. As a biological agent of terror it has advantages such as an *in-vitro* culture system, a large DNA genome amenable to genetic manipulation, a highly susceptible host, moderate to high infectivity, and, most importantly, person-to-person respiratory spread. In the United States, recommendations for routine smallpox vaccination ended in 1971. In 1976 routine smallpox vaccinations of health-care workers were also discontinued. In 1982 the only licensed manufacturer of vaccinia virus in the United States discontinued production and distribution to the civilian population. All military personnel continued to be vaccinated until 1990. Since then only select populations have been vaccinated. These include people working with vaccinia and monkeypox and health-care workers involved in clinical trials using recombinant vaccinia vectors. More recently, vaccinia trials have been conducted to determine the viability of the vaccine and the effective dose for restoring immunity. Recommendations have emerged for a national pre-event vaccination program [1–3].

This chapter is an attempt to combine the old with the new. Clinical experience with smallpox continues to diminish. It is in the descriptions and experience of physicians and health care workers who took care of thousands of cases in which clinical memory of the disease resides. Clinical research performed during smallpox epidemics helps bridge the gap between past and present animal models of vaccinia and variola major. An attempt has been made to cite that research. Finally, the descriptions of naturally occurring smallpox refer to a disease in which the inoculum and characteristics of the virus may be different from an intentional release during a bioterrorism event.

Bioterrorism Preparedness. Edited by Nancy Khardori
Copyright © 2006 WILEY-VCH Verlag GmbH & Co. KGaA, Weinheim
ISBN: 3-527-31235-8

4.2
History

Smallpox has caused epidemics throughout human history and is well described, because of its characteristic clinical presentation. It is possible smallpox was present in ancient Egypt from 1200 to 1100 BC [4]. Mummies with dome shaped vesicles in the epidermis similar to those of smallpox suggest the disease was present. Definitive descriptions of smallpox date from 4th and 7th-century China and India, respectively [5, 6]. Smallpox may have been introduced into France and Italy as early as 580 AD. Intranasal variation in China was described in 910 AD [7]. Most regions of the world have reported smallpox epidemics [8]. Initial introduction of smallpox into a non-immune population has resulted in high mortality rates. For example, an epidemic occurred in the village of Mine, a small island near Japan in 1795. There were 1200 cases of smallpox out of a population of 1400 with a case mortality rate of 38.3%. The first epidemic in Iceland in 1241 killed 20 000 of the total population of 70 000 and was followed by other epidemics in 1257 and 1291 [6]. By the 15th century, smallpox had become endemic in much of Europe. The first occurrence of smallpox in the Western Hemisphere occurred in 1507 after being imported from Spain. Settlers from France, the Netherlands, and Great Britain later imported smallpox into North America. The first epidemic occurred from 1617 to 1619 and killed many of the Indians. This epidemic pattern continued across North America. It has been reported that in 1738 smallpox killed half of the Cherokee tribe. Some of the 18th-century epidemics among the Native Americans seem to been initiated by deliberate action [9]. Sir Jeffrey Amherst, commander-in-chief of the British forces in North America, wrote to an officer in 1763 during the Pontiac rebellion and wondered if smallpox could be used to reduce the number of Native Americans. The plan was to inoculate blankets with smallpox and trade them with the Indians.

The settlers were not spared, however. Smallpox was usually imported on ships from Great Britain or, later, on African slave ships. There were numerous epidemics along the Eastern seaboard. Quarantine and isolation were practiced routinely. In 1721, variolation was introduced into the colonies followed by introduction of vaccination after 1799 [10]. Of note, Abraham Lincoln in 1863 was diagnosed with smallpox when the characteristic rash appeared two days after giving his Gettysburg address [11]. The last naturally occurring smallpox case was seen in Somalia in October 1977. Since that time there has been one unintentional releases of smallpox virus that occurred in a laboratory accident in Birmingham, UK, in 1978 [12–14]. In May 1980, the World Health Organization certified the world free of naturally occurring smallpox.

4.3
Virology

The *Poxviridae* comprise a family of DNA viruses that replicate in the cytoplasm of invertebrate and vertebrate cells (Table 4.1). The poxviruses are brick-shaped and

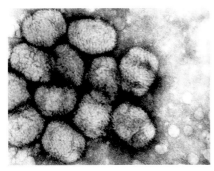

Fig. 4.1
Smallpox (variola) virus. Center for Disease
Control, Atlanta, Georgia.

enveloped, an important aspect of diagnosis by electron microscopy (EM) (Fig. 4.1). The virions are larger than those of other animal viruses and are the only animal viruses that can be seen by light microscopy. They carry linear, double-stranded, DNA and replicate in the cytoplasm of cells forming cytoplasmic inclusion bodies. Isolated vaccinia virions have been found to have RNA-synthetic activity [15]. These viruses also have the largest DNA genome; it varies from 130 to 230 kbp. Complete genome sequences have been reported for vaccinia, variola, and molluscum contagiosum virus [15–17]. The greatest variation in poxvirus genomes lies at each of the ends [18, 19] where non-essential and virulence genes may be located. Interestingly, several virus-encoded proteins that modulate the immune system and affect cell growth have been found. There are eight recognized genera of vertebrate poxviruses. Several of these orthopoxviruses infect humans, including variola, monkeypox , cowpox, and vaccinia sub species buffalopox virus. In addition, humans may become infected with pseudocowpox, Orf, seal parapoxvirus Tanapox, Yabapox, and molluscum contagiosum virus. Disease may vary from single or multiple localized skin lesions, after contact with infected animals, insects, or exposure to fomites, and after respiratory tract infection.

Tab. 4.1
Classification of vertebrate poxviruses (Chordopoxviridae).

Genus	Species
Avipoxvirus	Canary pox, fowlpox
Capripoxvirus	Goatpox, lump skin disease, sheeppox
Leporipoxvirus	Hare fibroma, myxoma
Molluscipoxpoxvirus	Molluscum contagiosum
Orthopoxvirus	Buffalopox, camelpox, cowpox, monkeypox, rabbitpox, racoonpox, tatera pox, variola, vaccinia
Parapoxvirus	Orf, pseudocow pox
Suipoxvirus	Swinepox
Yatapoxvirus	Tanapox, Yaba

Different types of smallpox disease have been observed in India and China. Smallpox (variola major) was regarded a disease with a high mortality rate. In 1904, a mild smallpox-like disease was described in South Africa. This was consequently termed variola minor. It was distinguished by a mortality rate of approximately 1%. This will not be discussed further, because it is unlikely it would be used in a bioterrorist attack.

4.4
Clinical Features and Classification

4.4.1
Rash and Prognosis

Rash is the hallmark of smallpox. As such, much emphasis has been placed on classification of the smallpox rash and its prognosticating significance. Previous classification schemes have emphasized rash density as being important [20, 21]. Later work recognized the type of rash as being more important than the density, with Raos classification subsequently being adopted, followed by modification by the World Health Organization (WHO) [6, 22, 23].

Table 4.2 outlines the classification of smallpox, including the prevalence and mortality rate, based on the type and confluence of the rash. The mortality rate from ordinary smallpox depends on the type and, less so, the confluence of the rash. Mortality rates can vary from 0 to 100%, depending on the smallpox type and vaccination status. Two points are important when discussing this classification system. The first is that the cases from Rao are a hospital series and may represent a sicker population. The second point is that measurement of vaccine status was based on a vaccine scar. Evidence of a vaccine scar may or may not correlate with successful vaccination, because a secondary bacterial infection of the vaccine site could masquerade as a successful vaccination. The history of vaccination is not always dependable, for the usual reasons of patient historical recollection and the lack of a potent vaccine. It was not until the advent of freeze-dried vaccine that potency could be guaranteed.

In an epidemic, knowledge of the type of rash may be important for prognosticating and triaging individual cases. The rash will be discussed further in Section 4.6.7.

4.5
The Stages of Smallpox

For the purposes of discussion and understanding of the disease, smallpox can be divided into three stages, i.e. the incubation, pre-eruptive, and eruptive stages.

These are not firm divisions and there may be overlap of signs and symptoms between each of the stages. In addition, it is not known what effect a large inoculum, which could theoretically occur in a bioterrorism event, might have on the timing of the different disease stages or the type of smallpox (ordinary vs. hemorrhagic).

Tab. 4.2
Clinical classification of variola major and mortality based on rash [23].

Type	Description	Vaccinated		Unvaccinated	
		Frequency (%)	Mortality (%)	Frequency (%)	Mortality (%)
Ordinary type	Total	70	3.2	88.8	30.2
	Confluent-confluent rash on face and forearms;	4.6	26.3	22.8	62
	Semiconfluent-confluent rash on face, discrete	7	8.4	23.9	37
	elsewhere;	58.4	0.7	42.1	9.3
	Discrete-areas of normal skin between pustules even on face				
Modified type	Total	25.3	–	2.1	–
	Confluent	0.1	–	0	–
	Semiconfluent	0.5	–	0.05	–
	Discrete	24.3	–	2	–
	Variola sine eruptione	0.4	–	0.05	–
Flat type	Total	1.3	66.7	6.7	96.5
	Pustules remained flat.	0.6	85.7	4.2	100
	Confluent	0.3	60	1.2	95.3
	Semiconfluent	0.4	42.1	1.3	95.6
	Discrete				
Hemorrhagic type	Total	3.4	93.9	2.4	96.4
	Early, with a purpuric rash; always fatal;	1.4	100	0.7	100
	Late, with hemorrhages into base of pustules; usually fatal.	2.0	89.8	1.7	96.8

4.5.1
Incubation Period

The incubation period, as defined from infection to fever onset is difficult to determine and depends on patients who have a discreet and limited contact with a source. In naturally acquired cases the range seemed to be 7 to 19 days. Most cases occurred between 10 and 14 days with a median of 12 days [24]. Jacobs, writing in a forward to a reprint of Rickettss book, stated that mild cases might have an incubation period of up to 3 weeks [21]. During this time, the subject is not felt to be infectious despite on-going viremia.

4.5.2
Pre-eruptive Stage

The pre-eruptive stage has been described as the time from onset of fever to the appearance of rash. The duration of the pre-eruptive stage has been reported to be highly variable [21]. The pre-eruptive stage was noted to be 2 to 3 days but could last as long as four to six [21, 23] and was accompanied by constitutional signs. Longer pre-eruptive stages were associated with hemorrhagic smallpox in which the rash might not appear. The appearance of a focal rash was accompanied by a decrease in fever with subjective feelings of improvement.

Fever

It was uncommon to have a smallpox case without a history of fever or constitutional symptoms. Fever was the most common symptom and was sudden in onset with temperatures between 38.5 °C and 40.5 °C. A typical history was of a patient who went to work in the morning and developed fever and headache and could not continue. It was stated by Rao that fever itself did not incapacitate the patient but the accompanying symptoms often did and felt this might be of importance in the differential diagnosis. During this time, the patient appeared toxic. The fever would decrease by the 2nd to 4th day and the patient would feel better. This would herald the onset of the eruptive stage. In Rao's series 60% of adults and 20% of children complained of rigors or chills. The chills were usually associated with the first fever [23].

Headache and Backache

Ninety percent of patients complained of a severe headache and backache (Table 4.3). Headache was often frontal and could precede fever onset although this was not absolute. Headache could also be a prominent complaint in children.

Tab. 4.3
Frequency of symptoms in the pre-eruptive phase of variola major [6].

Symptom	Variola major; 6942 cases (%)
Fever	100
Headache	90
Chills	60
Backache	90
Pharyngitis	15
Vomiting	50
Diarrhea	10
Delirium	15
Abdominal colic	13
Seizures	7

Delirium and Hallucinations

Fifteen percent of cases were described as being delirious, which may have been secondary to high fever in adults. There were occasional hallucinations. These signs disappeared on appearance of the rash.

Seizures

Seizures were seen in children and were usually associated with high fevers. Seizures were not felt to predict CNS involvement and recovery depended on the severity of the underlying smallpox.

Abdominal Symptoms

Abdominal pain occurred in 13% and 6% of adults and children, respectively [23]. Vomiting and abdominal colic sometimes occurred and could be mistaken for appendicitis.

Upper Respiratory Symptoms

It was stated by Rao that 15% of adults and children presented with a sore throat and a cough but not with sneezing, runny nose, or watering of the eyes [23].

Rash

In early writings there is reference to the appearance of a rash during the pre-eruptive phase that could lead an experienced clinician to the diagnosis of smallpox [21]. If these rashes appeared (<10%), it would be on the second or third day of the pre-eruptive stage, Ricketts describes two different rashes with the first being a pupuric or petechial rash. This rash was confined to the groin area and the upper third of the thighs and was thought to predict a more severe course, and sometimes, hemorrhagic smallpox. Rao stated that for the small number of subjects seen with this rash that progressed to an early hemorrhagic smallpox, variola elementary bodies (cytoplasmic inclusions) could be seen from the smears of these lesions [23]. A second rash was described by Ricketts as a rose or erythematous rash. It had a more general distribution, more predominant on the trunk although it could extend elsewhere. This rash was confused with scarlet fever or measles. In general, this rash was associated with a better prognosis.

It should be pointed out that not all authors have put as much emphasis on pre-eruptive rashes and have associated them with patients with a prior history of vaccination [22–24]. As such, there is no agreement on the importance of these rashes for smallpox diagnosis.

4.5.3
Eruptive Stage

Individual Lesion

In comparison with varicella, smallpox lesions are embedded in the skin but are also raised above the surface (Fig. 4.2). The lesions start as a small macule that is flat with the skin. The lesions evolve into papules that are solid feeling and may be pink. The papules begin to vacuolate and become larger over the next few days. The lesions had a barely discernable erythematous areola around them. Three different characteristics of smallpox lesions have been described. Some lesions were loculated and when pierced would retain fluid. At one time this was believed to be a means of distinguishing smallpox from other vesicular lesions but its value in cases that were diagnostically difficult was questionable [6]. The second characteristic of the lesion was how it felt. The lesions were described as shotty. Although they projected minimally above the skin they could be isolated between the thumb and forefinger and were described as having "hard foreign bodies embedded in the epidermis"[6]. Finally, individual lesions can be umbilicated, as seen in both herpesvirus and poxvirus infections. The term describes a central depression that is seen in the vesicle that may last into the pustular stage. The absorption of fluid and fibrinous threads lead to flattening of the lesion in preparation for eventual scabbing.

4.6
Ordinary Type

The words "ordinary smallpox" were used to describe the usual smallpox case. The descriptions "discrete", "semi-confluent", and "confluent" based on lesion density were associated with different mortality rates (Table 4.2). The most common smallpox rash in Rao's classification is the ordinary type seen in 70% and 88% of vaccinated and unvaccinated patients, respectively (Table 4.2). The ordinary type was divided into discrete, semi-confluent, and confluent. The discrete type was characterized by areas of normal skin on the face and on other parts of the body Figs 4.3 and 4.4). Confluent rash was characterized by a confluence of the rash on

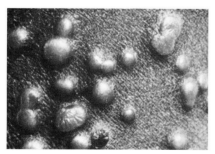

Fig. 4.2
Close-up of smallpox lesions on the thigh of a patient on day 6. Center for Disease Control, Atlanta, Georgia.

the face and forearms, with a lack of normal skin between lesions, whereas the semi-confluent type was defined as a confluent rash on the face but discrete on the body and forearms [6]. The mortality rate for discrete ordinary type in unvaccinated patients (9%) was lower than for either the semi-confluent type or the confluent ordinary type. Semiconfluent and confluent smallpox rash had mortality rates of 37% and 62%, respectively [23].

The first lesions appeared on mucous membranes and constituted the enanthem. These were tiny red lesions found on the oropharynx that evolved rapidly from macules followed by papules and then vesicles. The lesions were most prominent on the hard palate, tips, and edges of the tongue, and the tonsillar pillars. These lesions would appear approximately 24 h before the onset of skin rash and would break down and discharge virus into the oropharynx. The enanthem may be the reason some patients complained of sore throat.

The skin rash usually appeared 2 to 4 days after the onset of fever as a few macules on the face with a predilection for the forehead (herald spots). Lesions would then appear on the proximal extremities, followed by the trunk and, last, the distal portions of the extremities. This is termed centrifugal spread and would evolve rapidly. Rao wrote:

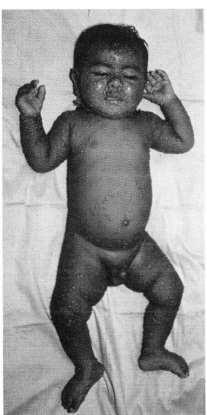

Fig. 4.3
Day 7. This is an example of ordinary smallpox. There is normal skin between lesions on the face and forearms. Note the centrifugal pattern with more extensive distribution on the face, arms, and shoulders [6].

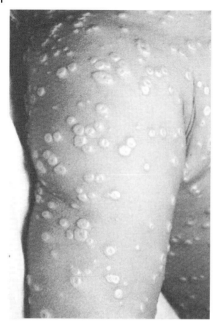

Fig. 4.4
Day 7 of rash in ordinary smallpox. Almost all lesions are at the same stage of development. Note the umbilicated lesions [6].

"The rash of smallpox has a characteristic centrifugal distribution pattern. It would almost seem that the patient, during the stage of viremia, had been fixed on a centrifuge with outstretched limbs and spun. The rash is more dense on the extremities than on the trunk; and, on the extremities, it is more dense on the distal parts than the proximal. On the face, it is more on the upper half than the lower half" [23].

The rash was not centrifugal in 10% of cases.

On the extremities, the rash was denser on the extensor than on the flexor surfaces. The rash would spare the hollow of the axilla and this was referred to as Rickett's sign and was usually positive except in a few instances. The rash was denser on the back than on the front of the trunk [21, 23]. On the front, it was denser on the chest than on the abdomen. This was referred to as Gaspirini's sign but was felt to be consistent in only 50% of cases [23]. The hands and feet were commonly involved, with the palms more affected than the soles [23]. Lesions on both the hands and the feet persisted even after scabbing had occurred on other parts of the body. This was because of the thick stratum corneum in these areas. Foot lesions would not protrude above the surface of the skin and could be seen as disk-like scabs called seeds [6].

In any given part of the body, all lesions would be at the same stage, although this was not absolute [21, 23]. There were no crops of lesions, as seen in chickenpox, but in smallpox the lesions may be seeded approximately the same time by the secondary viremia. That is not to say that on different parts of the body lesions may be in different stages because of to the centrifugal order of appearance. By day 2 after the appearance of the rash (day 5 of fever), the macules had evolved into what seemed to be papules. Histopathology showed these were early vesicles. By the fourth or fifth day after the appearance of rash (7th or 8th day from onset of fever), most lesions were vesicular. All the skin lesions were pustular by the 7th day after the appearance of the rash with the lesions reaching their maximum size by day 10 (day 10 to day 13 after onset of fever). The fluid absorbed and the central part of the lesion became hard followed by formation of a scab or crust. The crust separated after 22 to 27 days leaving a hypopigmented area. Lesions would persist on the palms of the hands and the soles of the feet after scabbing and resolution had occurred elsewhere. Fever would trend downward after the 4th or 5th day of rash but would rise again during the vesicular and pustular stages until scabbing had occurred.

4.6.1
Death

The death rate in ordinary smallpox varied, depending on the type, the confluency of the lesions, and previous vaccination status. Death usually occurred between days 12 and 18 after the onset of fever; this was referred to as the lethal period. What did patients die of and could modern medicine have an impact on the mortality rate from smallpox? Martin has reviewed pathology records during the last 200 years in an attempt to detail both the pathology and cause of death from ordinary smallpox disease [25]. The lungs were commonly involved, with degeneration of alveolar lining cells, bronchitis, and hyperemia. This picture was most consistent with viral pneumonia. There has been controversy about the role of bacterial pneumonia leading to death. In an earlier series, secondary bacterial pneumonia was felt to be a common cause of death but occurred after the lethal period during recovery when complications were more common [26]. Councilman found fatal bronchopneunomia to be the cause of death in one third of subjects who died in the lethal period. The role of bacterial pneumonia as a cause of death during the lethal period was questioned later [27–29]. Lillie believed most cases of bacterial pneumonia were actually viral pneumonia [28]. The series described by Bras is probably most relevant, because the data were was collected during the post-antibiotic era [29]. Fifteen percent of the patients in his series died of bacterial pneumonia. Even with better antibiotic therapy it would seem unlikely modern medicine would improve on this number, especially in a hospital setting in which bacterial superinfection may result from more antibiotic-resistant strains. As has been suggested, most deaths occurring in the lethal period, in which the role of bacterial pneumonia leading to death would seem to be less important, would not be affected by antibiotics. Martin concluded death was probably because of one or a combination of volume depletion, pneumonia, and renal failure. These were the result of the effect of viral replication and direct cytopathic effect. On the basis of timing of death and microbiological and pathological studies he concluded that bacterial sepsis or pneumonia were less likely. It would thus seem that the mortality rate from smallpox in the modern era would probably approach the numbers reported in the pre-antibiotic literature, especially in an outbreak that overwhelmed our current health-care system. This does not take into consideration any attempt to enhance the virulence of the smallpox strain used in a bioterrorist attack.

4.7
Modified-type

Modified-type smallpox has been defined as being accelerated in its clinical course compared with ordinary type and, in general, is expressed as a change in the number and severity of lesions. Patients would experience much of the pre-eruptive syndrome but the fever would become normal by the end of Day 6. The skin lesions may be more superficial, less in number and evolve more quickly, skipping the pustular stage and becoming crusted by Day 10. The most common cause of modified smallpox was prior vaccination. In the Rao series, 25 % of modified cases occurred in vaccinated patients whereas only 2 % were found in unvaccinated patients [23] (Table 4.2). There were no fatal cases associated with this rash.

4.7.1
Variola Sine Eruptione and Subclinical Infection

In well-vaccinated contacts of a household smallpox case a syndrome was observed in which a subject would have fever, backache, and headache that would subside within 48 h or less. The subjects never developed a rash. Variola major infection would be confirmed in the laboratory with an increase in complement fixation titer (6). In addition, studies have revealed recovery of variola major from the oropharynx in these subjects [30–32]. This presentation was exceedingly uncommon (Table 4.2).

Subclinical infection has been implicated by epidemiological studies in which increases in either complement fixation or gel precipitation antibodies were observed in populations exposed to smallpox [33, 34]. These finding have implications for immunity in a population and will be discussed further below.

4.8
Flat-type

Flat-type smallpox is so named because the lesions remain flat with the skin at the time of vesiculation compared with ordinary smallpox [6] (Fig. 4.5). An extensive enanthem was usually present on the tongue and hard palate. There was sometimes an extensive, severe enanthem on the rectal mucosa. The skin lesions matured slowly and, as the disease reached the papulovesicular stage, approximately 6 days after the onset of fever, a small depression was visible that became flat and buried in the skin by the 7th or 8th day. Many of these lesions would have hemorrhages at the base. The lesions contained little fluid and were not multi-locular, did not umbilicate, and felt soft and velvety to the touch [6]. The skin would peel from the lesion leaving a large denuded surface. The distribution of the lesions did not necessarily follow the centrifugal pattern seen in ordinary smallpox.

Patients with flat-type smallpox were febrile and toxic. Pulmonary edema was observed on the 7th or 8th day after the onset of fever. Several unvaccinated children developed acute stomach dilatation [23]. The color of the lesions would change to an ashen gray 24 to 48 h before death.

In the Rao series, flat type represented 6.7 % of total cases in unvaccinated individuals with most cases being in children (72 %). It was rare in vaccinated individuals. The prognosis was poor with an overall mortality rate of 96.5 % in unvaccinated individuals (Table 4.2) [23].

4.9
Hemorrhagic-type

Hemorrhagic-type smallpox was divided into early and late. Both were accompanied by a high mortality rate. In addition, both subtypes were different epidemiologically

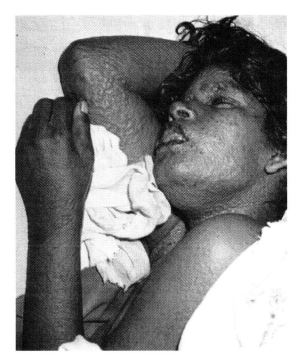

Fig. 4.5
Flat-type smallpox. The lesions are flat to the skin. The picture was taken before death [6].

from other smallpox types, because both were more common in adults and in vaccinated individuals [23, 35]. The data supporting hemorrhagic type being more common in vaccinated subjects was not universally seen [36]. The pathogenesis of hemorrhagic smallpox was not understood but was characterized by continual high grade viremia, unlike ordinary type, in which virus was rarely isolated after the appearance of the rash. Patients were also found to have thrombocytopenia and, in some studies, coagulation disorders [23]. Death was sudden and would usually occur before development of the smallpox rash. Hemorrhagic smallpox was uncommon but was also important for being commonly mistaken for other infectious diseases and its disproportionate occurrence in pregnant women.

4.10
Early Hemorrhagic-type

Early hemorrhagic-type smallpox presented with hemorrhages in the skin and mucous membranes before the development of rash. The disease had a sudden onset with severe headache and backache. A petichial rash may occur that may be confined to the groin area extending down to the upper third of the thighs [21]. Along with a petichial rash, the whole body may have a generalized erythema commencing on the second day of fever. Ricketts described the appearance of a subject with hemorrhagic smallpox with a characteristic face with immobile features, a lack of lines of expression, drooping eyelids, and profound prostration. He also stated the patients had an "unnatural clearness of the intellect" even until death [21]. Subconjunctival hemorrhages (60%) were common and could be accompanied by bleeding from the gums, hematemesis, hemoptysis, hematuria, melena, and epistaxis. Vaginal bleeding was common and could be seen in upwards of 70% of women (23). Death occurred around the 6th day of fever. It must be stressed that death was not because of bleeding; at autopsy the most consistent findings were evidence of heart failure and pulmonary edema [29]. If the subject survived longer, the superficial skin layers separated and formed large blebs containing serous or serosanguinous fluid. These ruptured leaving extensive areas of denuded skin.

When hemorrhagic-type smallpox occurred in adults, 88% of subjects were over the age of 14 years. It was found both in vaccinated and unvaccinated populations. Pregnant women were especially susceptible. Hemorrhagic-type smallpox was highly fatal but uncommon [23] (Table 4.2).

4.11
Late Hemorrhagic-type

This type was differentiated from early hemorrhagic type by the appearance of smallpox rash followed by hemorrhage in the lesions. The pre-eruptive stage was characterized by high fever (40 °C), severe constitutional signs which continued even after appearance of the rash. The lesions started as macules and became papules but only slowly progressed after that point. There would sometimes be hemorrhage in the base of the lesions giving them a flat appearance. Bleeding was similar to that of early hemorrhagic smallpox but less frequent. Subconjunctival hemorrhage (55%) and vaginal bleeding (60%) were still common [23]. Death would occur on the 8th to the 10th day. Mortality was high (Table 4.2.).

4.12
Complications

Rao states, "smallpox causes disfiguration, disability, deformity and death" [23]. Many different organ systems were affected by smallpox. Most complications were because of either direct viral invasion or secondary bacterial infection (Table 4.4).

Tab. 4.4
Complications of smallpox [6].

Organ system	Complication
Skin	Formation of boils secondary to bacterial infection. Sepsis. Pockmarks.
Eyes	Conjunctivitis. Corneal scarring.
Joints and bones	Arthritis. Viral invasion of the metaphyses of growing bones. Symmetrical elbow joint involvement
Respiratory system	Pulmonary edema. Pneumonitis. Cough was rare
Gastrointestinal system	Complications were uncommon except in flat-type smallpox.
Genitourinary system	Uncommon.
Central nervous system	Encephalitis was common occurring in about one in 500 cases. It usually started around the sixth day and recovery was slow but complete.

4.12.1
Skin

The skin was the most commonly effected, with scarring from direct viral invasion. Persistence of scars depended on lesion depth, which in turn depended on direct viral replication and not secondary bacterial infection. Initial scarring was universal but Rao stated that 30% of individuals diagnosed with smallpox did not have scars after 5 years [23]. The current suggestion is that scar formation is secondary to destruction of sebaceous glands, although virus has not been discovered there [29, 37].

4.12.1
Respiratory

Respiratory complications were common, especially in severe smallpox types. This varied from bronchitis, bacterial pneumonia, viral pneumonia, and pulmonary edema in hemorrhagic forms.

4.12.1
Gastrointestinal

In children with flat-type smallpox there was rectal mucosal involvement that could be quite severe. Before death, children might pass a cast of their rectal mucosa. In unvaccinated children, another complication could be acute stomach dilatation that would occur near day 12. It was a poor prognostic sign.

4.12.1
Neurological

Encephalitis was not uncommon, occurring in 1 in 500 patients [6, 23]. Patients had an altered sensorium with increased pressure in the CSF with a slight increase in cells but no biochemical changes. There was no nuchal rigidity or a Kernigs sign. Encephalopathy would occur on the 5th or 6th day and could continue until end of the third week of the illness. Recovery was usually complete but could be delayed for several months with gait returning to normal first, followed by speech.

4.12.1
Ophthalmic

In Raos series 1.7% of patients had eye involvement of some sort. Conjunctivitis was an uncommon manifestation. Keratitis was present in approximately 23% of patients with eye involvement. This was reported to be more common in malnourished children [22, 23]. Corneal ulcers occurred in 1% of all cases of smallpox and were responsible for 70% of eye complications. In 10% of cases, the ulcers were bilateral. Ulcers may start on or about the 10th day of the disease. In 40% of patients, this led to blindness in one eye [23].

4.12.1
Osteo-articular

A fairly common complication of smallpox infection was joint involvement. This occurred in 1.7% of cases and was more common in children. The elbow was the most commonly involved joint, often with bilateral disease [38]. Joint disease would manifest itself around the second to third week of illness. In children, involvement of the epiphysis with periostitis resembling bacterial osteomyelitis was present [39]. Most cases resolved without permanent sequelae but could lead to bone shortening, subluxations, and bone deformities.

4.13
Differential Diagnosis

Accurate clinical diagnosis of smallpox may be difficult early in the disease. Clinical diagnosis is based on the characteristic fever and constitutional signs, the order of appearance of rash, evolution of the lesions, and lesions on the same part of the body at the same stage. Differential diagnosis of smallpox is, however, the same as that of other diseases causing fever and rash (Table 4.5). These include monkeypox, chickenpox, measles, syphilis, erythema multiforme, drug eruptions, and Coxsackievirus infection. Some of the diseases listed are much less common than in previous decades. Confusion would arise if one of these common diseases occurred during a smallpox outbreak. Hemorrhagic-type smallpox has been misdiagnosed as meningococcal bacteremia or acute leukemia. By far the most common misdiagnoses are chickenpox (Table 4.6) and measles [6, 21,23, 40]. According to Ricketts, when measles is mistaken for smallpox it usually occurs in an adult. Since the onset of a monkeypox outbreak in the United States in 2003, this uncommon disease may need to be considered in the differential diagnosis [41–44]. Despite the availability of physicians with clinical experience, when smallpox was imported into the United Kingdom from distant colonies, misdiagnosis was not uncommon [45]. This often led to further spread of the disease.

Tab. 4.5
Differential diagnosis of fever and rash: causes of papulovesicular and maculovesicular eruptions [40].

Papulovesicular	Maculovesicular
Atypical measles (rubeola)	HIV
Acne	Adenovirus
Chickenpox	Arboviruses
Coxsackievirus (A16)	Atypical measles (rubeola)
Dermatitis herpetiformis	Cytomegalovirus
Drug eruptions	Drug eruptions
Eczema herpeticum	Epstein–Barr virus
Impetigo	Exanthem infectiosum (herpes virus 6)
Insect bites	Erythema infectiosum (parvovirus B19)
Molluscum contagiosum	German measles (rubella)

Papulovesicular	Maculovesicular
Monkeypox	Infectious mononucleosis
Papular urticaria	Measles (rubeola)
Pemphigus	Meningococcemia
Rickettsialpox	Kawasakis disease
Shingles	Mycoplasma pneumoniae
Yaws	Roseola infantum
Smallpox	Scalded skin syndrome (*Staphylococcus aureus*)
	Scarlet fever (*Streptococcus pyogenes*)
	Secondary syphilis
	Rat bite fever (*Streptobacillus moniliformis*)
	Reovirus
	Rocky mountain spotted fever
	Toxic erythemas
	Toxic shock syndrome
	Toxoplasmosis
	Typhus and tick fevers
	Typhoid
	Vaccine reactions

Tab. 4.6
Differential diagnosis of smallpox and chickenpox.

Symptoms	Smallpox	Chickenpox
Fever	2–4 days before rash	Same time as rash
Rash		
Appearance	Crop at same stage on the same body areas	Crop in several stages
Development	Slow	Fast
Distribution	Centrifugal.	Even distribution.
	More lesions on arms and legs	More lesions on trunk.
Palms and soles	Usually	Uncommon
Mortality	5–30%	Rare

At this time and without known release of smallpox virus, it is unlikely a patient presenting to physician with fever and a rash will have smallpox. The Center for Disease Control has supplied a case definition:

- Smallpox clinical case definition: An illness with acute onset of fever ≥101 °F (38.3 °C) followed by rash characterized by firm, deep seated vesicles or pustules in the same stage of development without apparent cause.
- Laboratory criteria for confirmation:
    – PCR identification of variola DNA in a clinical specimen, or
    – isolation of smallpox (variola) virus from a clinical specimen with confirmation by PCR [46].

In addition, as part of a process for screening subjects with fever and rash, the CDC has defined major and minor diagnostic criteria for smallpox. The major diagnostic criteria are:
- Febrile prodrome: occurring 1–4 days before rash onset: fever ≥ 101 °F (38.3 °C) and at least one of the following: prostration, headache, backache, chills, vomiting, or severe abdominal pain.
- Classic smallpox lesions: deep-seated, firm/hard, round, well-circumscribed vesicles or pustules; as they evolve, lesions may become umbilicated or confluent.
- Lesions in the same stage of development: on any one part of the body (e.g. the face, arms) all the lesions are all in the same stage of development (i.e., all are vesicles or all are pustules).

The minor diagnostic criteria are:
- centrifugal distribution – greatest concentration of lesions on face and distal extremities;
- first lesions on the oral mucosa/palate, face, or forearms;
- severity – patients seems toxic or moribund; and
- slow rash evolution – lesions evolved from macules to papules to pustules over days (each stage lasts 1–2 days) [46].

A high-risk subject would meet all three major smallpox criteria. A moderate-risk subject would have the febrile prodrome and meet one other major smallpox criterion, or the febrile prodrome plus all four minor smallpox criteria. In these diagnostic criteria, much emphasis is placed on the febrile prodrome. A question has been raised about the specificity of the smallpox prodrome, as it relates to helping exclude diagnosis of chickenpox which is widely believed not to have a prodrome. This has been questioned by one group who analyzed the presentation of chickenpox and found up to 7 to 17% of unvaccinated chickenpox case patients met the smallpox febrile prodrome criterion [47]. The CDC algorithm has been tested, however, and found to be helpful in guiding clinical and public health decisions [48].

4.14
Pathophysiology

Much of our understanding of smallpox pathogenesis is derived from studies of several animal models, including mousepox [49, 50]. This model has helped us understand viral spread within the host. Epidemiological evidence suggests variola virus enters through the respiratory tract. Variola has been shown to be capable of infection after aerosolization [51, 52]. It has been suggested that an outbreak of smallpox in Aralsk, Kazakhstan, was because of aerosolization during a bioweapons test resulting in virus being transmitted up to 15 km, but this is speculative [53] and the importance of aerosolization to other than close contacts remains controversial [54, 55]. The smallpox eradication program was based on person-to-person transmission being the most likely method of infection, and data suggest approximately 30% of susceptible individuals became infected during the era of endemic smallpox . Variola would enter the oro-nasopharynx or lung. It was felt the inoculum needed to initiate infection was low. Although the respiratory tract was the place of initial infection, a primary focus or lesion has not been found despite many searches [6]. After local replication, variola would seed the local or regional lymph nodes. The virus would continue to replicate and enter the bloodstream as cell-associated virus and spread to other body organs including marrow, lymph nodes, spleen, and liver. In mousepox, further replication occurred in the liver and spleen. In variola infection the liver shows evidence of infection but splenic involvement was less pronounced [25]. This stepwise process of viral replication was asymptomatic and encompassed the incubation period. The incubation period ended with the prodrome. Interestingly, one study showed that subjects excreted virus during the incubation period. Virus could be isolated by swabbing the oropharynx of subjects who were at high risk of infection as a result of caring for smallpox patients. Sarkar et al. used a single throat swab in subjects exposed to variola as a result of household contact. Thirty-four of 328 subjects who had been in contact with 52 smallpox cases grew variola [56]. Four of the 34 subjects with a positive culture subsequently developed clinical smallpox. This, however, was probably not an important means of spread of smallpox from person to person in comparison with the shedding of virus secondary to the enathem.

Poxviruses have adapted to survival in the host. Many of the adaptations depend on modulation of infection and inflammation [57–61]. They include intracellular serine protease inhibitor and double-stranded DNA binding protein that inhibit apoptosis and effect the interferon (IFN) response, respectively. It is also postulated that TNF, IL-18, chemokine, IFN-binding proteins reduce the specific and non-specific immune responses. Extrapolating from animal models and other human virus infections, what seems to be important in recovery from smallpox is the variola-specific CD8+ CTL response. The development of a.CD4+ Th1 response seems to be important in affecting the CD8+ response to poxviruses [62]. Some of the poxvirus immunomodulatory proteins block Th1-mediated responses, attempting to affect the cytokine environment that favors generation of virus-specific CTL [57, 59–61]. The effect of upsetting the Th1/Th2 balance was shown in an experiment in which the IL-4 gene was incorporated into ectromelia virus, resulting in increased virulence [63]. In addition, myxoma virus infection results in down-regulation of Class I MHC surface molecules that would hamper CD8+ CTL responses [64].

4.15
Laboratory Diagnosis

Laboratory confirmation of smallpox is especially important because of the implications for infection control. Specimens should be collected by someone who has been vaccinated and is properly protected by gown, gloves, and a N-95 fitted respirator. It may be necessary to open lesions with the point of the scalpel. The fluid can be harvested on a cotton swab or scabs can be picked off with forceps. Specimens should be deposited in a sealed container (Vacutainer, screw top tube, etc) and taped shut. The tube should be put in another water-tight container and sent to the appropriate state or federal laboratory [65]. Laboratory examination requires a BL-4 facility. Diagnosis of smallpox can be suggested by electron microscopy. Definitive identification requires growth in tissue culture, chorioallantoic egg membranes, and PCR confirmation [66]. A pustular lesion could, however, be tested for VZV and HSV by means of direct fluorescent antibody tests. Both tests are very sensitive and specific for their respective viruses. A lesion that is negative for both viruses in the appropriate clinical setting should raise clinical suspicion of smallpox. Many laboratories can perform direct fluorescent microscopy on skin lesions within 2 h.

4.16
Postexposure Infection Control

As soon as smallpox is diagnosed, all suspected patients, household contacts, and other face-to-face contacts should be vaccinated. If possible, patients and contacts should be isolated in the home. Vaccination seems to be effective within the first three to four days after exposure and provides some protection against infection and significant protection against death [6]. People should be vaccinated irrespective of previous vaccination history.

Hospital transmission of smallpox has been recognized for several hundred years. Transmission occurs via droplets and aerosols. Infection may occur at a distance when smallpox has spread through the ventilation system. It has been recommended that all hospital workers be vaccinated during an outbreak. For those who cannot receive the vaccine, vaccinia immunoglobulin can be given. In a small outbreak, patients admitted to the hospital should be confined to negative-pressure rooms equipped with high-efficiency air-particulate filtration. All laundry and waste should be placed in biohazard bags and autoclaved [65].

4.17
Vaccination and Immunity

Although vaccination has proven to be an effective means of protection against smallpox, it is evident when looking at past literature that vaccination was not always protective and even individuals that had been vaccinated several times could still manifest smallpox. In the United States, Dryvax , a live virus preparation of vaccinia virus is licensed for use [67] After the September 11th, 2001, terrorist attacks, the preparation was recognized in case a mass vaccination campaign became necessary. Vaccine trials showed the viability of the stored vaccinia virus. The vaccine preparation could be diluted as much as 1 to 10 and still give a high rate of conversion. A more recent paper has demonstrated that 90% of pre-vaccinated recipients could achieve a successful "take" on receiving vaccine diluted 1:3.2 (10^7 pfu mL^{-1}) [1, 2]. These trials are now less important because more vaccine has been produced. The Advisory Committee on Immunization Practices (ACIP) had decided not to recommend mass vaccination of US citizens but a more conservative strategy of vaccinating health-care workers and first responders has been adapted [68, 69]. In the event of intentional release of variola virus, groups that would be vaccinated include:

- persons exposed to the initial release of virus;
- persons who had face-to-face, household, or close-proximity contact (<6.5 ft or 2 m) with a confirmed or suspected smallpox patient at any time from the onset of the patients fever until all scabs had separated;
- personnel involved in direct medical or public-health evaluation, care, or transport of confirmed or suspected smallpox patients;
- laboratory person involved in the collection or processing of clinical specimens from confirmed or suspected smallpox patients; and
- other persons with increased likelihood of contact with infectious materials from a smallpox patient (e.g. personnel responsible for medical waste disposal, linen disposal or disinfection, and room disinfection, in a facility where smallpox patients are present).

There are no absolute contraindications to smallpox vaccination after a release of virus. High-risk groups for vaccine-related side effects are also at high risk of death from smallpox [70].

New data have emerged regarding the side-effect profile of mass vaccination. Cassimatis et al. reported on two confirmed and fifty probable cases of post-smallpox vaccine myopericarditis [71]. Between December 2002 and June 2003, 468 000 US military personnel were vaccinated. Non-cardiac vaccine side effects were less than or equal to historical levels except for myopericarditis. Cardiac involvement may range from asymptomatic non-specific ST-T wave abnormalities to a constellation of symptoms including chest pain, shortness of breath, dry cough, palpitations, syncope, paroxysmal nocturnal dyspnea, and others. Treatment usually consists in administration of non-steroidal anti-inflammatory agents, rest, and control of heart failure symptoms. When some populations have been vaccinated, old and new vaccine-related morbidities have surfaced [72–75]. Other vaccination complications have been well documented [76].

A ring vaccination strategy has been recommended in the event of a terrorist attack, because of the low probability of intentional smallpox release and concern about vaccine-related morbidity and mortality in a nation-wide vaccination program.

An important question remains about residual immunity in vaccine recipients. Because a sufficient number of doses is now available, the question of residual immunity is less important. Residual immunity does, however, have an effect on

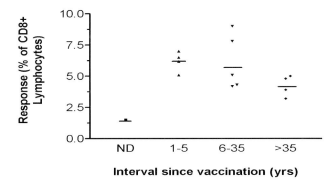

Fig. 4.6
Response to Vaccinia virus by CD8+ T lymphocytes at different
times after vaccination. From Ref. [79].

how well the variola might spread in the US. Some older data suggest that the
effects of vaccination may last for several decades [6, 77]. Table 4.7 lists data from
an outbreak in Liverpool, UK, and the effect of vaccination. There was a marked
difference between the incidence of disease in vaccinated and nonvaccinated
subjects. Protection decreases as the interval after vaccination increases [6]. Table 4.8, also, shows that protection is still evident after 10 years [77]. These data are,
unfortunately, difficult to interpret secondary to intercurrent infection in the
populations studied. Other *in-vitro* data have revealed that vaccinia-specific CD4+
CD8+ CTL may be recovered from vaccine recipients after several decades [62, 78,
79] but it is not known whether that would translate into protection (Fig. 4.6).

Tab. 4.7
Residual vaccine immunity – 1902–1903 outbreak [6].

Age group	Vaccination in infancy	Number of cases	Case totality (%)
0–4	+	7	0
	–	55	45
5–14	+	96	0
	–	57	10.5
15–29	+	436	0.7
	–	72	13.9
30–49	+	349	3.7
	–	24	54.2
> 50	+	55	5.5
	–	12	50

Tab. 4.8
Residual vaccine immunity – secondary attack rates in vaccinated and unvaccinated subjects in West Pakistan [77].

Status	Number of contacts	Smallpox cases (%)
No pre-exposure vaccine		
Vaccinated within 10 days	16	12 (75)
Not vaccinated within 10 days	27	26 (96)
Total	43	38 (88)
Pre- exposure vaccination		
<10 years	115	5 (4)
>10 years	65	8 (12)
Total	180	13 (7)
Previous smallpox	27	0 (0)
Total all contacts	250	51 (20)

4.18
Antiviral Treatment

The concept of antiviral therapy for treatment and prophylaxis is limited because of a lack of clinical trials. Vaccinia immune globulin (VIG) is available through the CDC but is currently reserved for vaccine-induced disease. There has, however, been substantial clinical experience with these preparations [80–82]. Several compounds have been shown to have promising activity inhibiting smallpox virus both *in vitro* and *in vivo*. Ribavirin has been shown to inhibit variola major *in vitro* [83]. More recently, cidofovir has been shown to have activity against smallpox virus both *in vitro* and *in vivo* [83–88]. An oral analogue is being developed [84, 89, 90]. Cidofovir has a long intracellular half-life but has been shown to be toxic to the renal tubules. It is necessary to give probenecid and to hydrate the patient before giving the drug. Whether this would be different with oral analogues is not known.

New research has suggested that the Abl family of tyrosine kinases may have a role in poxvirus replication. Reeves et al. showed that cell-associated vaccinia virions use the Abl and Src-family tyrosine kinases in actin motility. In addition, the Abl family was important in the release of cell-associated virions from the cell [91]. New inhibitors of the Abl tyrosine kinases were found to reduce viral titers by five orders of magnitude.

4.19
Summary

The literature on variola and smallpox is extensive. This chapter is an attempt to give more information about the clinical presentation of smallpox as a disease that is passed from person to person. The risk of intentional or unintentional release of variola remains difficult to quantify for health-care workers. It is also difficult to visualize exactly what a bioterrorist attack with variola would resemble. Would it be a mass release of virus as an aerosol infection of 1000, 5000, or more people at once? Trying to model epidemics, and calculating the spread of a virus on the basis of limited outbreaks, do not seem to adequately take into account what would happen in a mass casualty scenario in which the ability to isolate patients and perform contact tracing and ring vaccination would be overwhelmed [45, 92]. Would smallpox be a different disease in an individual receiving 1000–10 000-fold more virus than in a natural infection? Could the virus be engineered for greater virulence? These are questions that cannot be answered by a practicing clinician. By understanding more about smallpox, we are better able to confront it.

References

1 Frey, S. E., Couch, R. B., Tacket, C. O., et al. **2002** Clinical responses to undiluted and diluted smallpox vaccine. [see comment]. *N Engl J Med*; 346 (17):1265–1274.

2 Frey, S. E., Newman, F. K., Cruz, J., et al.**2002** Dose-related effects of smallpox vaccine. *N Engl J Med*; 346 (17):1275–1280.

3 Cohen, J. and Enserink M. **2002** Public health. Rough-and-tumble behind Bush's smallpox policy. *Science*; 298 **5602**:2312–2316.

4 Anonymous. **1991** Earliest known victim of small pox? *Nurs RSA Verpleging*; 6 (11/12):40.

5 Needham, J. **1980** China and the origins of immunology. *Eastern Horizon*; 19 (1):6–12.

6 Fenner, F., Henderson, D., Arita, I., et al. **1988** Smallpox and its eradication. Geneva: World Health Organization.

7 Buck, C. **2003** Smallpox inoculation – should we credit Chinese medicine? *Complement Ther Med*; 11 (3):201–202.

8 Thein, M. M., Goh, L. G. and Phua, K. H. **1988** The smallpox story: from variolation to victory. *Asia–Pacific J Public Health*; 2 (3):203–210.

9 Patterson, K. B. and Runge, T. **2002** Smallpox and the Native American. *Am J Med Sci*; 323 (4):216–222.

10 Hopkins, D. R. **1977** Benjamin Waterhouse (1754–1846), the "Jenner of America". *Am J Trop Med Hyg*; 26 (5 Pt 2 Suppl):1060–1064.

11 Basler, R. P. **1972** Did President Lincoln give the smallpox to William H. Johnson? Huntington Library Quarterly; 35:279–284.

12 Anonymous. **1979** WHO had serious doubts on safety in smallpox lab. *Nature*; 277 **5692**:79–80.

13 Anonymous. **1979** All safety nets failed, says Shooter. *Nature*; 277 **5692**:78–79.

14 Anonymous. **1979** UK smallpox research could continue at Porton. *Nature*; 277 **5692**:77–78.

15 Moss, B. **2001** Poxviridae: The viruses and their replication. In: Knipe D, Howely P, eds. *Field Virology*. 4th edn. Philadelphia: Lippincott, Williams, and Wilkins; pp. 2849–2883.

16 Shchelkunov, S. N., Totmenin, A. V., Babkin, I. V., et al. **2001**Human monkeypox and smallpox viruses: genomic comparison. *FEBS Lett*; 509 (1):66–70.

17 Shchelkunov, S. N., Totmenin, A. V., Safronov, P. F., et al. **2002** Multiple genetic differences between the monkeypox and variola viruses. *Doklady Biochem Biophys*; 384:143–147.

18 Muller, H., Wittek, K., Schaffner, W., et al. **1977** Comparison of five poxviruse genomes by analysis with restriction endonucleases Hind III, BamI, and EcoRI. *J Gen Virol*; 38:135–147.

19 Mackett, M. and Archard, L. C. **1979** Conservation and variation in Orthopoxvirus genome structure. *J Gen Virol*; 45:683–701.

20 Ricketts, T. **1893** A classification of cases of smallpox by the numerical severity of the eruption. London: McCorquodale.

21 Ricketts, T.**1966** The diagnosis of smallpox. Reprinted by U. S. Department of Health, Education, and Welfare, Public Health Service, London: Cassell.

22 Dixon, C. **1962** Smallpox. London: J and A Churchhill.

23 Rao, A. R. **1972** Smallpox. 1st edn. Bombay: Kothari Book Depot.

24 Fenner, F. **1980** The global eradication of smallpox. *Med J Aust*; 1(10): 455–455.

25 Martin, D. B. **2002** The cause of death in smallpox: an examination of the pathology record. *Mil Med*; 167(7):546–551.

26 Councilman, W., Magrath, G., Brinkerhoff W. **1904** The pathological anatomy and histology of variola. *J Med Res*; 11(12–135).

27 Perkins, R. and Pay, G. **1903** Studies on the etiology and pathology of variola. *J Med Res*; 10(163–179).

28 Lillie, R. **1930** Smallpox and vaccinia: The pathologic histology. *Arch Pathol*; 10:241–291.

29 Bras, G. **1952** The morbid anatomy of smallpox. *Doc Med Geogr Trop*; 4:303–351.

30 Verlinde, J. and Tongeren, H. A. **1952** Isolation of smallpox virus from the nasopharynx of patients with variola sine eruptione. Antonie van Leeuwenhoek. *J Microbio Serol*; 18:109–112.

31 Marennikova, S. Gurvich E. B. and Ma, Y. **1963** Laboratory diagnosis of smallpox and similar viral diseases by means of tissue culture methods. I. Sensitivity of tissue culture methods in the detection of variola virus. *Acta Virol*; 7:124–130.

32 Shelukhina, E. M., Marennikova, S. S., Maltseva, N. N., et al. **1973** Results of a virological study of smallpox convalescents and contacts: short communication. *J Hyg Epidemiol Microbiol Immunol*; 17(3):266–271.

33 Heiner, G. G., Fatima, N., Daniel RW, et al. **1971** A study of inapparent infection in smallpox. *Am J Epidemiol*; 94(3):252–268.

34 Rao, A. R., Sukumar, M. S., Kamalakshi, S., et al. **1970** Precipitation in gel test in diagnosis of smallpox. *Indian J Med Res*; 58(3):271–282.

35 Mazumder, D. N. **1973** Smallpox on the increase. *J Indian Med Assoc*; 60(8):303.

36 Sarkar, J. K., Mitra, A. C. and Chakravarty, M. S. **1972** Relationship of clinical severity, antibody level, and previous vaccination state in smallpox. *Trans Roy Soc Trop Med Hyg*; 66(5):789–792.

37 Regan, T. D. and Norton, S. A. **2004** The scarring mechanism of smallpox. *J Am Acad Dermatol*; 50(4):591–594.

38 Gupta, S. K. and Srivastava, T. P. **1973** Roentgen features of skeletal involvement in small pox. *Aust Radiol*; 17(2):205–211.

39 Cunha, B. A. **2004** Smallpox and measles: historical aspects and clinical differentiation. *Infect Dis Clin N Am*; 18(1):79–100.

40 Breman, J. G. and Henderson, D. A. **2002** Diagnosis and management of smallpox.[see comment]. *N Engl J Med*; 346(17):1300–1308.

41 Anonymous. **2003** Update: multistate outbreak of monkeypox – Illinois, Indiana, Kansas, Missouri, Ohio, and Wisconsin, 2003. *MMWR*; 52(26):616–618.

42 Stephenson, J. **2003** Monkeypox outbreak a reminder of emerging infections vulnerabilities. *JAMA*; 290(1):23–24.

43 Ligon, B. L. **2004** Monkeypox: a review of the history and emergence in the Western hemisphere. *Semin Pediatr Infect Dis*; 15(4):280–287.

44 Di Giulio, D. B. and Eckburg, P. B. **2004** Human monkeypox: an emerging zoonosis.[see comment][erratum appears in Lancet Infect Dis. 2004 Apr; 4(4):251]. *Lancet Infect Dis*; 4(1):15–25.

45 Pennington, H. **2003** Smallpox and bioterrorism. *Bull World Health Organ*; 81(10):762–767.

46 Anonymous. **2002** Bioterrorism watch. Smallpox or chickenpox? How to make the diagnosis. *Ed Manag*; 14(1):Suppl 3–4.

47 Moore, Z., Seward, J., Watson, B., et al. **2004** Chickenpox to smallpox: The use of the febrile prodrome as a distinguishing characteristic. *Clin Infect Dis*; 39:1810–1817.

48 Seward, J., Galil, K., Damon, I., et al. **2004** Development and experience with an algorithm to evaluate suspected smallpox cases in the United States, 2002–2004. *Clin Infect Dis*; 39:1477–1483.

49 Fenner, F. **1948** The clinical features and pathogenesis of mouse-pox (infectious ectromelia of mice). *J Pathol Bacteriol*; 60:529–552.

50 Fenner, F. **1948** The pathogenesis of acute exanthems. *Lancet*; 2:915–920.

51 Noble, J., Jr. and Rich, J. A. **1969** Transmission of smallpox by contact and by aerosol routes in Macaca irus. *Bull World Health Organ*; 40(2):279–286.

52 Wehrle, P. F., Posch, J., Richter, K. H., et al. **1970** An airborne outbreak of smallpox in a German hospital and its significance with respect to other recent outbreaks in Europe. *Bull World Health Organ*; 43(5):669–679.

53 Zelicoff, A. P. **2003** An epidemiological analysis of the 1971 smallpox outbreak in Aralsk, Kazakhstan. *Crit Rev Microbiol*; 29(2):97–108.

54 Meikeljohn, G., Lempe, C., Downie, A. W., et al. **1961** Air sampling to recover variola virus in the environment of a smallpox hospital. *Bull World Health Organ*; 25:63–67.

55 Downie, A. W., Meikeljohn, G., St. Vincent, L., et al. **1965** The recovery of smallpox from patients and their environment in a smallpox hospital. *Bull World Health Organ*; 33:615–622.

56 Sarkar, J. K., Mitra, A. C. and Mukherjee, M. K. **1973** Serial isolation of variola virus in the throat of household contacts of smallpox cases. *Bull Calcutta Sch Med*; 21(2):21–22.

57 Smith, S. and Kotwal, G. **2002** Immune response to poxvirus infections in various animals. *Crit Rev Microbiol*; 28:149–185.

58 Harte, M. T., Haga, I. R., Maloney, G., et al **2003** The poxvirus protein A52R targets Toll-like receptor signaling complexes to suppress host defense. *J Exp Med*; 197(3):343–351.

59 Belyakov, I. M., Earl, P., Dzutsev, A., et al. **2003** Shared modes of protection against poxvirus infection by attenuated and conventional smallpox vaccine viruses. Proceedings; 100(16):9458–9463.

60 Seet, B. T., Johnston, J. B., Brunetti, C. R., et al. **2003** Poxviruses and immune evasion. *Annu Rev Immunol*; 21:377–423.

61 Bray, M. and Buller, M. **2004** Looking back at smallpox. *Clin Infect Dis*; 38(6):882–889.

62 Amara, R. R., Nigam, P., Sharma, S., et al. **2004** Long-lived poxvirus immunity, robust CD4 help, and better persistence of CD4 than CD8 T cells. *J Virol*; 78(8):3811–3816.

63 Jackson, R., Ramsay, A., Christensen, C., et al. **2001** Expression of mouse interleukin-4 by a recombinant ectromelia virus suppresses cytolytic lymphocyte responses and overcomes genetic resistance to mousepox. *J Virol*; 75:1205–1210.

64 Nash, P., Barrett, J., Cao, J. X., et al. **1999** Immunomodulation by viruses: the myxoma virus story. *Immunol Rev*; 168:103–120.

65 Henderson, D. A., Inglesby, T. V., Bartlett, J. G., et al. **1999** Smallpox as a biological weapon: medical and public health management. Working Group on Civilian Biodefense. *JAMA*; 281(22):2127–2137.

66 Kulesh, D. A., Baker, R. O., Loveless, B. M., et al. **2004** Smallpox and pan-orthopox virus detection by real-time 3'-minor groove binder TaqMan assays on the Roche LightCycler and the Cepheid smart Cycler platforms. *J Clin Microbiol*; 42(2):601–609.

67 Rotz, L. D., Dotson, D. A., Damon, I. K., et al. **2001** Advisory Committee on Immunization P. Vaccinia (smallpox) vaccine: recommendations of the Advisory Committee on Immunization Practices (ACIP), *MMWR*; 50(RR-10):1–25; quiz CE1–7.

68 Cohen, J. and Enserink, M. **2002** Rough-and-tumble behind Bush's smallpox policy. *Science*; 298:2312–2316.

69 Anonymous. **2003** Recommendations for using smallpox vaccine in a pre-event vaccination program. *MMWR*; 52((RR-7)):1–16.

70 Goldstein, J. A., Neff, J. M., Lane, J. M., et al. **1975** Smallpox vaccination reactions, prophylaxis, and therapy of complications. *Pediatrics*; 55(3):342–347.

71 Cassimatis, D. C., Atwood, J. E., Engler, R. M., et al. **2004** Smallpox vaccination and myopericarditis: a clinical review.[see comment]. *J Am Coll Cardiol*; 43(9):1503–1510.

72 Grabenstein, J. D. and Winkenwerder, W, Jr. **2003** US military smallpox vaccination program experience.[see comment]. *JAMA*; 289(24):3278–3282.

73 Halsell, J. S., Riddle, J. R., Atwood, J. E., et al. **2003** Myopericarditis following smallpox vaccination among vaccinia-naive US military personnel.[see comment]. *JAMA*; 289(24):3283–3289.

74 Talbot, T. R., Stapleton, J. T., Brady, R. C., et al. **2004** Vaccination success rate and reaction profile with diluted and undiluted smallpox vaccine: a randomized controlled trial.[erratum appears in *JAMA*. 2004 Nov 24; 292(20):2470]. *JAMA*; 292(10):1205–1212.

75 Oh, R. C. **2005** Folliculitis after smallpox vaccination: a report of two cases. *Mil Med*; 170(2):133–136.

76 Lane, J. M., Ruben, F. L., Neff, J. M., et al. **1970** Complications of smallpox vaccination, 1968: results of ten statewide surveys. *J Infect Dis* 122(4):303–309.

77 Mack, T. M., Thomas, D. B., Ali, A., et al. **1972** Epidemiology of smallpox in West Pakistan. I. Acquired immunity and the distribution of disease. *Am J Epidemiol*; 95(2):157–168.

78 Demkowicz, W. E., Jr., Littaua, R. A., Wang, J., et al. **1996** Human cytotoxic T-cell memory: Long lived responses to vaccinia virus. *J Virol*; 70(4):2627–2631.

79 Frelinger, J. A. and Garba, M. L. **2002** Responses to smallpox vaccine. [comment]. *N Engl J Med*; 347(9):689–690; author reply 689–690.

80 Sharp, J. C. and Fletcher, W. B. **1973** Experience of anti-vaccinia immuno-globulin in the United Kingdom. *Lancet*; 1**7804**:656–659.

81 Feery, B. J. **1976** The efficacy of vaccinial immune globulin. A 15-year study. *Vox Sanguinis*; 31(1 SUPPL):68–76.

82 Hopkins, R. and Lane, M. **2004** Clinical efficacy of intramuscular vaccinia immune globulin: A literature review. *Clin Infect Dis*; 39:819–826.

83 De Clercq, E. **2001** Vaccinia virus inhibitors as a paradigm for the chemotherapy of poxvirus infections. *Clin Microbiol Rev*; 14:382–397.

84 Bray M, Martinez M, Smee DF, Kefauver D, Thompson E, Huggins JW. Cidofovir protects mice against lethal aerosol or intranasal cowpox virus challenge. *J Infect Dis* 2000; 181(1):10–19.

85 Bronze MS, Greenfield RA. Therapeutic options for diseases due to potential viral agents of bioterrorism. *Curr Opin Invest Drugs* 2003; 4(2):172–178.

86 Neyts J, Clercq ED. Therapy and short-term prophylaxis of poxvirus infections: historical background and perspectives. *Antiviral Res* 2003; 57(1/2):25–33.

87 Bray M, Roy CJ. Antiviral prophylaxis of smallpox. *J Antimicrob Chemother* 2004; 54(1):1–5.

88 Prichard MN, Kern ER. Orthopoxvirus targets for the development of antiviral therapies. *Curr Drug Targets – Infect Disorders* 2005; 5(1):17–28.

89 Ciesla SL, Trahan J, Wan WB, Beadle JR, Aldern KA, Painter GR, et al. Esterification of cidofovir with alkoxyalkanols increases oral bioavailability and diminishes drug accumulation in kidney. *Antivir Res* 2003; 59(3):163–171.

90 Keith KA, Wan WB, Ciesla SL, Beadle JR, Hostetler KY, Kern ER. Inhibitory activity of alkoxyalkyl and alkyl esters of cidofovir and cyclic cidofovir against orthopoxvirus replication *in vitro*. *Antimicrob Agents Chemother* 2004; 48(5):1869–1871.

91 Reeves PM, Bommarius B, Lebeis S, McNulty S, Christensen J, Swimm A, et al. Disabling poxvirus pathogenesis by inhibition of Abl-family tyrosine kinases.[see comment]. *Nature Med* 2005; 11(7):731–739.

92 Kretzschmar M, van den Hof S, Wallinga J, van Wijngaarden J. Ring vaccination and smallpox control. *Emerg Infect Dis* 2004; 10(5):832–841.

5
Anthrax – Bacteriology, Clinical Presentations, and Management

Nancy Khardori

5.1
Historical Background

The earliest known historical writings about anthrax from Egypt and Mesopotamia date back to 5000 B.C. [1]. The Book of Exodus in the Bible describes the Fifth Plague (1491 BC) as killing the cattle of the Egyptians [2, 3]. The disease described as the Fifth Plague seems to have been anthrax. The Sixth Plague described in the same book may have been outbreaks of anthrax in humans [4]. Descriptions of anthrax involving animals and humans appear in the early literature of Hindus, Romans, and Greeks. The name "anthrax" derives from the Greek root for "coal" describing the black eschar in the cutaneous form of the disease. The first pandemic in Europe, known as the "Black Bane", occurred in the 17th century and caused many human and animal deaths. The disease in humans was later described as the "malignant pustule." The first epidemic of anthrax in the US occurred in the early 18th century. Outbreaks of occupational cutaneous and respiratory anthrax were reported from industrial European countries in the mid 1800 s. Cutaneous anthrax was shown to be caused by handling wool, hair, and hides. Respiratory disease (inhalational anthrax) was caused by processes that created an aerosol, for example cording wool, and was given the name "Woolsorter's Disease."

Delefond in 1838 described the microscopic appearance of *Bacillus anthracis* and Devain described its infectivity in 1868. Koch's extensive microbiological work on the organism and the disease resulted in the discovery of single organism etiology for infectious diseases. Anthrax became the first human infectious disease attributed to a specific microbial agent when Koch showed it to fulfill his "postulates" in 1877. Pasteur field tested in sheep the attenuated anthrax spore vaccine in 1881. In the early 1900 s the reduced use of potentially contaminated imported animal products and improved industrial and animal husbandry hygiene led to a steady decrease in the annual number of anthrax cases in the developed countries. The development of an animal vaccine from the spore suspension of an avirulent,

Bioterrorism Preparedness. Edited by Nancy Khardori
Copyright © 2006 WILEY-VCH Verlag GmbH & Co. KGaA, Weinheim
ISBN: 3-527-31235-8

nonencapsulated live strain was reported by Sterne in 1939. This is the animal vaccine currently in use. Both live attenuated and killed vaccines have been developed for human use [5]. The cell free vaccine for humans, a sterile filtrate of cultures from an avirulent, nonencapsulated strain that elaborates protective antigen was licensed in the US in 1970. China and Russia use a live spore-based vaccine. The largest recorded outbreak of human anthrax, and probably the largest among animals, occurred in Zimbabwe in 1978–1980 during the time of civil war [6, 7]. Not much has been written about this outbreak outside of Africa. The human disease followed an unprecedented epidemic in cattle and caused an estimated 10,000 cases and 151 deaths. An accidental outbreak of anthrax occurred in Sverdlovsk (now Ekaterinburg, Russia) in 1979 down wind of a Soviet military microbiology facility.

5.2
Epidemiology

Anthrax is predominantly a zoonotic disease of herbivores in arid and semiarid climates acquired while grazing in areas of high soil contamination [8]. Herbivores can also acquire the disease from the bite of certain flies. Vultures may be involved in the mechanical spread of the disease in the environment. Few, if any, warm-blooded species are entirely immune to anthrax. Before the availability of the animal vaccine and antimicrobial therapy, anthrax was one of the foremost causes of uncontrolled mortality in cattle, sheep, goats, horses, and pigs, worldwide [9]. Cattle are highly susceptible to systemic disease, shed large numbers of the organism (that contaminate the environment) and die within 1–2 days of infection. Animal vaccination programs have reduced animal mortality substantially. Humans are incidental hosts and almost invariably contract anthrax directly or indirectly from animals. Anthrax spores continue to be documented in soil samples from throughout the world. The factors that favor sporulation of *B. anthracis* include:
1. alkaline soils (pH > 6.0);
2. high nitrogen levels in the soil owing to decaying vegetation;
3. alternating periods of rain and drought; and
4. temperatures above 15.5 °C.

Areas with such conditions are known as "incubator areas". Other sources of animal anthrax include excreta and saliva from dead or decaying animals, blood-sucking flies, carrion eaters, wool and hair wastes, imported bonemeal and vegetable proteins, contaminated commercial animal feed, cleanings used in fertilizers, and tannery effluents.

The disease is still enzootic in most countries of Africa and Asia, several European countries, areas of the American continent, and some areas of Australia. Sporadic cases are reported from other parts of Australia. Anthrax in domestic animals is seen primarily in the tropics and subtropics, including India, Pakistan, Africa, and South America. In African wildlife, in which control measures including vaccination are irrelevant or difficult to accomplish, anthrax remains a major cause of uncontrolled mortality in herbivores [3]. Anthrax spores can resist environmental degradation and survive for many decades. Some environmental conditions seem to lead to heavily and persistently contaminated soil in areas called "anthrax districts" or "anthrax zones." In the US , "anthrax zones" lie parallel to the cattle drive trails of the 1800s [10]. In the US, anthrax is unusual as a result of vaccination of animals and preventive industrial hygiene measures. Anthrax in animals occurs occasionally in the Midwest and Western US. Spores in the soil are ingested by cattle or other herbivores and germinate into the vegetative form in the spleen and lymph nodes resulting in bacteremia and hemorrhage as a terminal event [2]. The vegetative forms from the infected animals are deposited in the soil, followed by sporulation and continuation of the cycle of infection. It is believed that cycling of the bacteria through herbivores who die and release bacteria into the soil and surrounding water sources explain local soil contamination. Altered patterns of grazing or distribution of spores in the soil may be responsible for seasonal variation in the incidence of disease. The recent decontamination of Gruinard Island (off the coast of Scotland) which was deliberately contaminated during World War II, has shown that the spores can be eliminated from a defined area by costly but simple techniques [11].

The World Health Organization (WHO) maintains the data on the global epidemiology of anthrax in animals and humans available on a comprehensive web site [12]. The organization also provides a comprehensive resource on guidelines for control of anthrax in animals and humans [9].

5.3
Microbiology and Genetics

Bacillus anthracis belongs to the genus *Bacillus*. The spore size of *B. anthracis* is approximately 1 μm. They grow readily at 37 °C on ordinary laboratory media with non-hemolytic "curled hair" colony morphology and a "jointed bamboo-rod" cellular appearance [13, 14]. Spores also germinate readily in blood or tissues of animals or human hosts, environments rich in amino acids, nucleosides, and glucose. The vegetative forms are nonflagellated and large (1–8 μm in length and 1–1.5 μm in breadth). They are aerobic or facultatively anaerobic and grow readily on sheep blood agar. After overnight growth at 35 °C, a well isolated colony is 2–5 mm in diameter, white or gray-white, tenacious, flat with a "medusa head" appearance (Fig. 5.1). The organism is nonmotile, nonhemolytic, catalase-positive, and forms a central or subterminal spore. It is lysed by γ-phage and under natural conditions is susceptible to penicillin.

Commercially available API test strips and fluorescent antibody staining can be used for further identification. The CDC has provided detailed guidelines on

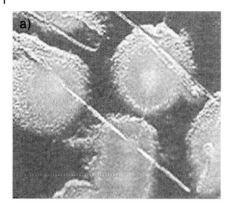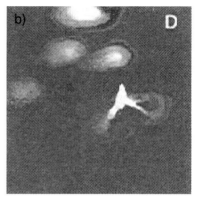

Fig. 5.1
(a) *B. anthracis*, colony on SBA. (b) "Sticky" consistency of *B. anthracis* colony on SBA (from CDC, Atlanta, Georgia).

preliminary testing and referral to the public health laboratories for confirmation [15]. Polymerase chain reaction (PCR) and γ-phage lysis are used to confirm the microbiological diagnosis. Testing for capsule formation on bicarbonate agar under anaerobic conditions is an additional test for identification of *B. anthracis*. The polypeptide capsule consisting of poly-d-glutamic acid can be visualized by India ink staining. Clinical samples, e.g. pus or tissue specimens from patients suspected of having anthrax should be stained with Gram stain to reveal gram-positive bacilli. Polychrome methylene blue can be used to show the polypeptide capsule [3].The rapidly multiplying vegetative forms of *B. anthracis* form spores only after local nutrients are exhausted such as when infected body fluids are exposed to ambient air. The Northern Arizona University (NAU) laboratory has used amplified fragment length polymorphism (AFLP) DNA analysis to examine 1200 strains identified around the world over the years. *B. anthracis* is a highly monomorphic species with genes from different isolates having more than 99% nucleotide sequence identity [16]. Multilocus variable number tandem repeat (VNTR or MLVA) markers are the most accurate strain-typing tool available. MLVA has been performed on half of the 1200 known strains of *B. anthracis*. This is a highly precise method for genetic finger printing. It takes the NAU laboratory approximately 12 h to analyze an anthrax sample. VNTR analysis of the Florida, New York, and Washington, D. C. isolates recovered from the 2001 letter attacks suggested they were from the same source, a strain originally isolated from a dead cow in Texas in 1981. This strain, designated Ames, has been used in the US Defensive Biological Weapons Program at the US Army Medical Research Institute (USAMRIID) in Fort Detrick, Maryland. It was also provided to other research laboratories in the US and Europe. The Institute for Genomic Research (TIGR) in Rockville, Maryland, has identified a nearly complete genome sequence of a *Bacillus anthracis* Ames isolate (the Porton isolate) that lacks the virulence plasmids. The sequences from the Florida isolate were aligned to the Porton chromosome and the previously sequenced *B. anthracis* plasmids. A comparative genomics approach was used to identify polymorphic sites in the genome. The Florida isolate revealed 60 new markers including single nucleotide polymorphisms (SNP_S). Comparison of the two genomes revealed four high quality SNP_S between the two chromosomes and seven differences among different preparations of the reference genome. When tested on a collection of *B. anthracis* isolates, these markers were found to divide the samples into distinct families. Such genome-based analysis of microbial agents provides a powerful new tool for investigation of infectious disease outbreaks and microbial forensics [16].

5.4
Virulence Factors and Pathogenesis

The two unique virulence factors in the fully virulent strains of *B. anthracis* are:
1. a poly-d-glutamic acid capsule that inhibits phagocytosis; and
2. a tripartite toxin [17].

The capsule is produced by virulent strains of *B. anthracis in vivo* and when grown on media containing serum or bicarbonate, or both, and incubated anaerobically in a CO_2-enriched atmosphere. The capsule is encoded by the plasmid pXO2 [18]. The existence of a toxin produced by *B. anthracis* was demonstrated in 1955 in a guinea-pig model – injection of sterile plasma from infected guinea pigs resulted in local edema and death. The toxin was later shown to have three separate components. The individual toxin components have no known biological effects. The three-component toxin consisting of edema factor, protective antigen, and lethal factor is coded by the plasmid pX01 [19]. The presence of the plasmid and the regulatory gene at atxA which resides on that plasmid are required for virulence. The sequencing and organization of pX01 and pX02 have already been analyzed [20–22]. Protective antigen (PA) named for its role in protective immunity binds to cell-surface receptors to produce an uptake system used by both edema factor and lethal factor to gain intracellular access. Protective antigen also inhibits phagocytosis of anthrax bacilli by polymorphonuclear cells. The cellular receptor, the anthrax toxin receptor (ATR), for PA was identified recently [23]. The receptor is a Type 1 membrane protein with an extracellular Von Willebrand factor A domain that binds directly to PA. A soluble version of this domain can protect cells from the action of the toxin. The edema factor (EF) is a calmodulin-dependent adenylate cyclase that converts ATP to cyclic adenosine monophosphate (cAMP). Increased levels of intracellular cAMP result in dysregulation of water and ions, including calcium. Such dysregulation may lead to edema in a manner analogous to loss of water into the intestinal lumen caused by cholera toxin. Edema factor also inhibits phagocytosis of *B. anthracis* by polymorphonuclear cells. The structure of EF was determined by X-ray crystallography in 2002 [24]. The enzyme was shown to be activated only after binding to calmodulin and undergoing a conformational change. The lethal factor (LF) is a zinc-dependent metallopeptidase that cleaves members of the mitogen-activated protein kinase kinase (MAPKK) leading to inhibition of one or more signaling pathways. The crystal structure of LF showed it to be comprised of four domains. Domain 1 binds to the protective antigen [25].

"Edema toxin" (edema factor and protective antigen) and "lethal toxin" (lethal factor and protective antigen) resemble the A–B enzyme-binding structures characteristic of many well-studied bacterial toxins. The lethal toxin LT was recently shown to severely impair the function of dendritic cells and induce profound impairment of antigen-specific T and B-cell immunity [26]. Earlier, LT had been shown to suppress rather than induce proinflammatory cytokine production in macrophages [27]. Suppression was seen with very low levels of LT and involves inhibition of transcription of cytokine messenger RNA.

The pathogenesis of fatal anthrax is not completely known but is thought to be caused by anthrax toxin. The binding of PA to its cellular receptor (ATR) results in its cleavage by a cellular protease into 20-kDa and 63-kDa portions. A heptamer is formed by seven copies of the 63-kDa PA and remains bound to the ATR in a doughnut-shaped complex. Next LF or EF (maximum of three copies) or a combination of both bind to the PA heptamer. The complex is then internalized and enters the intracellular endosome. The low pH of the endosome enables the LF and EF to cross the endosome membrane into the cytosol [28]. In the cytosol, the LF and EF trigger their toxic effects, which include immune system evasion and cell damage. Sellman et al. have shown that a mutant PA molecule could form a part of the doughnut like normal PA but could not disrupt the membrane pore, preventing escape of EF and LF [29]. The pathogenic effects of anthrax toxin have previously been attributed to macrophage-mediated cytokine release (TNF-λ and IL-1); toxin-induced lysis of infected macrophages, and septic shock resulting in death [30]. More recent work in a murine model has shown that lethal toxin kills by a mechanism that is cytokine-independent [31, 32]. The death in these mice involved striking tissue hypoxia and liver necrosis. The exact mechanism of this damage remains to be defined.

5.5
Human Anthrax – Clinical Manifestations

5.5.1
Cutaneous Anthrax

The most common naturally occurring form of human disease caused by *B. anthracis* is cutaneous anthrax, also known as "Malignant Pustule". An estimated 2000 cases are reported annually worldwide [33]. In the US, 224 cases of cutaneous anthrax were reported over the 50 year period 1944–1994 [34]. One case was reported in the year 2000. After the anthrax attacks of 2001, 11 confirmed or probable cases of cutaneous anthrax occurred in the US. Cutaneous anthrax typically occurs after an exposure to anthrax-infected animals and follows the deposition of the organism into the skin. The lesions occur mostly on exposed areas of the body, such as arms, hands, neck, and face. Although previous cuts or abrasions increase susceptibility, in the only published case of cutaneous anthrax resulting from the 2001 US attacks the patient did not have prior visible cuts or abrasions at or around the site of the lesions. The incubation period in these cases ranged from one to ten days with a mean of five days based on estimated date of exposure to *B. anthracis*-contaminated letters [14]. Cutaneous anthrax occurred only as late as twelve days after the original release of aerosol. After germination of the anthrax spores in skin tissue, toxin production leads to localized edema. This is followed by a pruritic macule or papule which enlarges to become a round ulcer by the second day. The ulcer is then surrounded by 1 to 3-mm vesicles with clear or serosanguinous fluid that reveals Gram-positive bacilli on Gram stain. The development of a painless, depressed, black eschar associated with extensive local edema follows. As the eschar dries, it loosens and falls off in the next 1–2 weeks. The local lesion can be associated with lymphangitis and painful lymphadenitis. The infectious causes in the differential diagnosis include plague, tularemia, scrub typhus, rickettsial spotted fevers, rat bite fever, and ecthyma gangrenosum. Different forms of vasculitis and arachnid bites can resemble the eschar caused by anthrax. Two excellent resources of information on clinical diagnosis of cutaneous anthrax have recently become available [35, 36]. Antibiotic therapy reduces edema and the likelihood of systemic disease but does not change the course of the skin lesion. The mortality rate in untreated cases has been reported to be 20% but with antibiotic treatment deaths from cutaneous anthrax are rare.

5.5.2
Gastrointestinal Anthrax

Although uncommon, outbreaks of gastrointestinal anthrax in Asia and Africa continue to be reported. No culture-proven cases of gastrointestinal anthrax were reported from the US between 1900 and 2000. Between 1960 and 1974, more than 100 cases of the intestinal form of anthrax were reported from the Bekka Valley of Lebanon [37]. Two recent events in which anthrax seems to have caused intestinal disease were reported in the US [38, 39]. In August 2000, five members of a Minnesota family ate meat from a cow that was later shown to be infected with *B. anthracis*. The meat was reported to have been well cooked. Two of the family members developed a self-limited gastrointestinal illness within 48 h of eating the meat. Antibiotic prophylaxis with ciprofloxacin and anthrax vaccine were given even though the symptoms had resolved. The second event was reported in one of the patients who died from the 2001 anthrax attack. The patient had nausea, vomiting, and abdominal pain. Abdominal CT scan showed findings consistent with necrotizing enteritis. At autopsy, 2500 mL of ascitic fluid with necrotizing infection, hemorrhage, and Gram-positive bacilli in the ileum were seen. Gastrointestinal anthrax occurs after ingestion of insufficiently cooked contaminated meat [14]. Direct gastrointestinal instillation of *B. anthracis* spores did not produce experimental infection in primates [40]. Because of the rapid transit time in the gastrointestinal tract, germination of spores may be unlikely. It is more likely that most cases result from ingestion of large numbers of vegetative forms of *B. anthracis* present in poorly cooked infected meat. Not much is known about the risks from direct contamination of food and water.

Gastrointestinal anthrax occurs in two clinical forms:

1. the oral-pharyngeal form results in oral or esophageal ulcers and regional lymphadenopathy, edema, and septicemia [41];
2. intestinal anthrax, the more common form, manifests as primary lesions predominantly in the terminal ileum or cecum [12]. The disease presents initially as nausea, vomiting, and malaise and progresses rapidly to bloody diarrhea, acute abdomen, and sepsis. Massive ascites may occur. The late stage of the disease may appear similar to the systemic disease or sepsis syndrome secondary to cutaneous or inhalational anthrax [14]. Thick yellow ascites, segmental bowel edema, and mesenteric lymphadenopathy, particularly in the ileocecal area, are seen on exploratory laparotomy. Because of the difficulty of early diagnosis, the mortality rate from gastrointestinal anthrax is high. On the basis of experience from the Lebanese outbreak, the authors suggest that intestinal anthrax should be treated with medical therapy including, initially, intravenous antibiotics. If a response is not seen quickly, wide surgical resection of the areas of bowel involved, extending into apparently uninvolved tissue, should be performed. The ascitic fluid should be drained continuously, because of potential post-operative re-accumulation. The overall estimated mortality rate of intestinal anthrax is 28 to 60% [38].

5.5.3
Inhalational Anthrax

"Inhalational anthrax" refers to the route of acquisition of infection. This form of the disease (approx. 5% of cases reported worldwide) leads to the most serious morbidity of the natural form and would be expected to do the same if *B. anthracis* were used as an aerosolized biological weapon [14]. Inhalational anthrax is always a medical emergency. In the US only 18 cases of inhalational anthrax were reported between 1900 and 2000; the mortality rate was 89%. Most of these cases occurred before the development of critical care units and before the availability of antibiotics. The known risk groups, including goat hair mill, wool, and tannery workers and laboratory workers, were affected [43]. Recent analysis of data obtained after accidental release of *B. anthracis* in Sverdlovsk suggests there may have been as many as 250 cases with 100 deaths [44]. Cases occurred from 2 to 43 days after exposure. In experimental infections of monkeys, fatal disease occurred 58–98 days after exposure. Mediastinal nodes of one monkey showed viable spores 100 days after exposure [45]. The anthrax attacks of 2001 in the US resulted in eleven cases of inhalational anthrax, five of whom died (case fatality rate 45 %) [2, 14]. The median period from presumed time of exposure to onset of symptoms was 4 days (range 4–6 days). Patients with fulminant illness before antibiotic administration all died. Clinical presentation of inhalational anthrax has been described as starting with a spectrum of nonspecific symptoms (fever, dyspnea, cough, headache, vomiting, chills, weakness, abdominal pain, and chest pain) and laboratory abnormalities. This stage lasts from hours to a few days. With or without a brief period of apparent recovery, the second stage develops abruptly with fever, dyspnea, diaphoresis, and shock. Stridor is seen in some cases because of mediastinal lymphadenopathy and expansion of the mediastinum. Kyriacou et al. compared the clinical presentation of 47 historical cases (including eleven cases of bioterrorism-related anthrax) with 376 controls with community-acquired pneumonia or influenza-like illness [46]. Nausea, vomiting, pallor or cyanosis, diaphoresis, altered mental status, and raised hematocrit were seen more frequently in the anthrax patients than in either of the control groups. The most accurate predictor of anthrax was X-ray evidence of mediastinal widening or pleural effusion with a sensitivity of 100 % and a specificity of 71.8 compared with community-acquired pneumonia and 95.6 % compared with influenza-like illness. Hemorrhagic meningitis with signs of meningitis, delirium, and obtundation develop in 50 % of patients. Later, second stage hypotension and cyanosis progress rapidly with death sometimes occurring within hours. The clinical presentation and laboratory findings in the 2001 US inhalational anthrax cases are described in detail by Inglesby et al. [14].

Inhalational anthrax develops after deposition of spore bearing particles of 1 to 5 μm in alveolar spaces. Some of the spores are lysed and destroyed in the macrophages. Spores surviving in the macrophages are transported via the lymphatics to mediastinal lymph nodes. Germination of the spores in the mediastinal nodes may take up to 60 days. Disease follows rapidly when germination occurs because of the release of toxins by the replicating vegetative bacteria. Edema, hemorrhage, and necrosis are the pathological hallmarks. Toxin-mediated effects and death can occur even if the bloodstream is sterilized with antibiotics. On the basis of experimental data from primates, the LD_{50} for humans (lethal dose sufficient to kill 50 % of persons exposed) has been estimated to be 2500 to 55000 inhaled anthrax spores [45]. The LD_{50} for aerosolized anthrax spores in cynomolgus monkeys has been reported to be 8000 colony-forming units [47]. In such experiments fatality was 20–80 %. The minimum infective dose of anthrax for humans would be important in events in which residual spores remain in the environment. Determination of LD_{10} (dose required to kill 10 %) or LD_1 (dose required to kill 1 %) would require a much larger number of experimental monkeys but would be much lower than the LD_{50}. LD_1 is, however, more relevant to large potentially exposed populations. On the basis of the discordant LD_{50} values of 4100 spores and 8000 spores, it has been extrapolated that LD_{10} is 50 or 98, LD_5 is 14 or 28, LD_2 is 4 or 7, and LD_1 is 1 or 3 spores. This LD_1 value explains the rare sporadic cases described in people with minimal contact with a known contaminated environment. The results from this extrapolation are not, however, in accordance with older observations that unimmunized workers in wool mills could inhale hundreds of spores daily without developing disease [48]. Host variables, underlying lung disease, and differences in the virulence of anthrax strains may determine the susceptibility of humans to anthrax. Increased susceptibility of mice to anthrax has been linked to a gene on chromosome 11 for a poorly understood protein called KifIC. Mice with a mutated version of the gene were protected.

5.5.4
Hemorrhagic Meningoencephalitis

Neurological complications of anthrax can occur with any of the three forms of primary anthrax infections [49, 50]. Meningoencephalitis should be regarded as a fourth type of initial clinical presentation of anthrax and should prompt a search for inhalational anthrax and potential bioterrorism [2]. The autopsy reports from Sverdlovsk cases revealed hemorrhagic meningitis in 50% of patients. The first recognized case in the US 2001 anthrax attack was a patient in Florida who presented with hemorrhagic meningitis [51]. Historically the fatality is 95%, even for treated cases. Other neurological complications include parenchymal brain hemorrhages, hematomas, vasculitis, subarachnoid hemorrhage, and cerebral edema.

5.5.5
Microbiological Diagnosis

The local or state health department, local hospital epidemiologist, and local or state health laboratory must be immediately notified on first suspicion of an anthrax illness. Microbiological detection of organisms resembling *B. anthracis* may be the means of initial detection of an outbreak. In advanced infection with a high bacterial burden the bacilli may be visible on Gram-staining of unspun peripheral blood (Fig. 5.2). Gram-staining of the peripheral blood is not a procedure performed routinely in diagnostic laboratories, however. The most useful microbiological test is the standard blood culture (Fig. 5.3). It is very

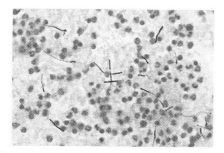

Fig. 5.2
Gram stain of peripheral blood buffy coat from a patient with inhalational anthrax in 2001. (Reproduced with permission from Borio et al. (2002). Death Due to Inhalational Anthrax in Bioterrorism, Guidelines for Medical and Public Health Management. Edited by Henderson. Inglesby and Toole. AMA Press, Chicago, IL, USA.)

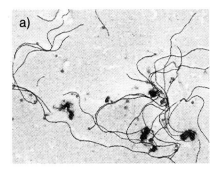
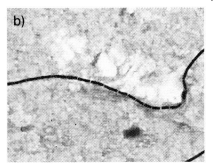

Fig. 5.3

Gram stain of blood culture media from a patient with inhalational anthrax in 2001. (Reproduced with permission from Borio et al. (2002). Death Due to Inhalational Anthrax in Bioterrorism, Guidelines for Medical and Public Health Management. Edited by Henderson. Inglesby and Toole. AMA Press, Chicago, IL, USA.)

important to obtain blood cultures before initiation of antibiotics. Growth in blood cultures is usually observed after 6 to 24 h and may be sterilized after only 1 or 2 doses of antibiotics. It takes approximately 24 h after growth to identify a Bacillus species from a blood culture. Unless the laboratory has been alerted to the possibility of anthrax, the Bacillus species are not identified further. Isolation of Bacillus species in the US most often indicates growth of *Bacillus cereus*. If diagnosis of anthrax is entertained, biochemical testing and review of colony morphology could provide a preliminary diagnosis 12 to 24 h after inoculation of the cultures.

After preliminary identification, the isolate should be sent promptly to one of the facilities in the Laboratory Response Network (LRN). The LRN in the US has been established by collaboration between the CDC and the Association of Public Health Laboratories [52]. Diagnosis of bioweapons pathogens can currently be made in 81 clinical laboratories in the LRN. Confirmatory tests used in these laboratories include immunohistochemical staining, enzyme-linked immunosorbent assay for protective antigen, gamma phage lysis, and polymerase chain reaction assay. The Mayo–Roche rapid anthrax test, a rapid-cycle real-time polymerase chain reaction, is commercially available [53]. This assay has the advantage of combining amplification of target DNA with detection of amplicons in the same closed reaction vessel. Real-time PCR formats coupled with rapid thermocycling in a single system enable a much shorter turn-around time for results and sensitivity exceeding that of standard culture-based assays. Espy et al. reported detection of Vaccinia virus, Herpes simplex virus, Varicella-zoster virus, and *Bacillus anthracis* DNA by light-cycle polymerase chain reaction after autoclaving of the specimens [54]. The standard autoclaving procedures eliminated the infectivity of the pathogens. Such tests offer the possibility of immediate diagnosis by specific qualified laboratories without Biosafety Level 4 facilities. Polymerase chain reaction for amplification of specific virulence plasmid markers in different anthrax strains may soon become available [55]. Because of the frequent lack of a pneumonic process in inhalational anthrax, respiratory secretions are unlikely to reveal the organism by Gram-stain and culture. In only one of the 2001 inhalational anthrax cases in the US was the organism revealed by Gram-stain of the sputum. In suspected cases of cutaneous anthrax, a Gram-stain and culture of the vesicular fluid should be obtained. If the patient is already on antibiotics and/or Gram-stain of the fluid is negative, punch biopsy of the skin should be sent to a laboratory able to perform immunohistochemical staining or polymerase chain reaction assays. Blood cultures should be obtained from patients with cutaneous anthrax before starting antibiotics. The presence of Gram-positive bacilli in the CSF in a patient with a compatible clinical illness should alert the clinician and the laboratory staff to diagnosis of anthrax. This is how the index case of inhalational anthrax in the 2001 attacks was identified [51]. Postmortem findings of thoracic hemorrhagic necrotizing lymphadenitis and hemorrhagic necrotizing mediastinitis and/or hemorrhagic meningitis in a case with unexplained death should strongly favor the diagnosis of anthrax.

The predictive value of the nasal swab testing for diagnosing inhalational anthrax in humans is unknown and untested. The CDC does not recommend nasal swab as a diagnostic specimen. Epidemiologically, potentially exposed persons who have positive nasal swab tests for *B. anthracis* should receive a course of post-exposure antibiotic prophylaxis. Negative results should not be used to rule out infection in a patient, however.

5.5.6
Immunological Tests and Serological Diagnosis

The major immunogenic proteins of *B. anthracis* are the capsular antigens and the anthrax toxin components [55]. A fourfold rise in antibody titers by specific enzyme-linked immunosorbent assays (ELISAs) against these components is diagnostic of past infection or vaccination. Antibody titers to protective antigen and capsular components are the most reliable indicators. Other immunological tests that have been studied include enzyme-linked immunoelectro-transfer blotting and indirect microhemagglutination. Although useful for epidemiological purposes, these tests should not be used for diagnosis of anthrax in the acutely ill patient. Antibodies to toxin or capsular components cannot be detected until late in the course of the disease and no increase in antibody titer is seen in treated patients. Immunological tests to detect exotoxins in blood during systemic infection are also unreliable for diagnosis. A commercially produced chemical extract of an attenuated strain of *B. anthracis* is available for diagnosis of acute and previous infections with *B. anthracis* [56]. When used for skin testing, this extract was positive in 82% of cases 1–3 days after onset of symptoms and 99% of cases by the end of the fourth week.

5.5.7
Antimicrobial Therapy and Post-exposure Prophylaxis

In addition to supportive therapy for acutely ill patients, all persons in high-risk groups for potential exposure to *B. anthracis* should receive antimicrobial therapy as soon as possible while awaiting the results of laboratory tests [14]. No antibiotic regimen, including those commonly used for presumptive treatment of severe sepsis, has been studied for treatment of inhalational anthrax in humans. Most naturally occurring strains of *B. anthracis* are resistant to extended spectrum cephalosporins and are sensitive to penicillin. Historically, penicillin has been the preferred antibiotic for treatment of anthrax. An inducible β-lactamase in addition to a constitutive cephalosporinase has been described recently. These strains are highly susceptible to penicillin *in vitro* (MIC less than 0.06 µg mL^{-1}). It is possible that a large bacterial burden of such strains can overcome the in-vivo efficacy of penicillin. Doxycycline has been proven to be efficacious for anthrax in studies on monkeys [57]. Other members of the tetracycline class are suitable alternatives. All fluoroquinolones have in-vitro activity against *B. anthracis*. Although no human studies are available, animal models suggest excellent efficacy for ciprofloxacin [57, 58]. Penicillin, doxycycline, and ciprofloxacin are currently approved antimicrobial agents for anthrax [59]. Because of the theoretical concern over the inducible β-lactamase, CDC did not advise treatment with penicillin or amoxicillin as monotherapy. On the basis of in-vitro activity, efficacy in the monkey model, and FDA approval ciprofloxacin or doxycycline should be regarded as the standards [14]. Other antibiotics with in-vitro activity against *B. anthracis* include clindamycin, rifampin, imipenem, aminoglycosides, chloramphenicol, vancomycin, cefazolin, macrolides, and linezolid. A *B. anthracis* strain engineered to be resistant to tetracycline and penicillin has been reported. More recently, in-vitro resistance to ofloxacin (from the fluoroquinolone class) has been reported to develop after subculturing and multiple cell passages in an isolate of the Sterne strain of *B. anthracis* [60]. Data from 2001 anthrax attacks suggested that persons with inhalational anthrax treated intravenously with two or more antibiotics with in-vitro activity against *B. anthracis* had a greater chance of survival [61]. Although data are limited, combination antimicrobial therapy is a reasonable therapeutic approach for life-threatening anthrax. This may be particularly useful in cases with central nervous system involvement because of consideration of CNS penetration. Ciprofloxacin in combination with chloramphenicol, rifampin, or penicillin is recommended when meningitis is a consideration [14]. Addition of clindamycin has also been recommended on the basis of the theoretical benefit of diminishing bacterial toxin production [62].

Initial intravenous antibiotic therapy (ciprofloxacin or doxycycline) then oral therapy with the same agents is recommended in the contained casualty setting. In a mass causality setting intravenous therapy and/or combination therapy may no longer be practical. For such events, oral therapy with ciprofloxacin or doxycycline is recommended for adults for therapy or post-exposure prophylaxis. For children and pregnant women ciprofloxacin or amoxicillin are recommended.

There are no FDA-approved antibiotic regimens for post-exposure chemoprophylaxis. Antibiotic therapy or post-exposure prophylaxis should be continued for at least 60 days after exposure, because of the possibility of delayed germination of spores. Treatment of cutaneous anthrax also should be for 60 days after exposure, because of presumed concomitant inhalational exposure in the setting of a potential bioterrorism event. Depending on epidemiological circumstances, public health officials will need to provide guidelines about populations requiring post-exposure prophylaxis. The high-risk groups should be instructed to report any flu-like or febrile illness immediately. They should be evaluated for the need to initiate treatment for possible inhalational anthrax, because of uncertainties about the duration of the latency of the spores.

5.5.8
Emerging/Investigational Therapies

Passive immunotherapy using plasma from vaccinated horses was the only available anthrax treatment in the pre-antibiotic era and is still used in Russia and China [63]. There are no scientific data about its efficacy in humans. In animal studies antibody therapy has been shown to be effective only when given before anthrax infection. The Centers for Disease Control and Prevention (CDC) and other federal agencies have been discussing the use of plasma from military personnel vaccinated against anthrax to provide preformed antibiotics against the anthrax toxin. Such "antitoxin" therapy would be used only as an adjunct to antibiotics and only for patients not responding to antibiotics. The US Army Medical Research Institute of Infectious Diseases (USAMRIID) in collaboration with the CDC and the National Institutes of Health (NIH) is conducting animal experiments to determine the amount of plasma that would be needed, and its efficacy. The current plasma supply collected from military personnel is small. The second larger batch from vaccinated volunteers would add modestly to these supplies some of which will be used in animal studies. A potent antibody against the protective antigen (PA) subunit of anthrax toxin has been isolated *in vitro* using an *Escherichia coli* expression system to create a library of antibodies displayed in phage [64]. Selected antibodies from this library were shown to bind to PA with high affinity. The antibody (IH) with the highest affinity for PA prevented anthrax toxin from binding to its receptor on cultured alveolar macrophages and protected rats against a lethal challenge with the anthrax toxin. Iverson and Georgiou have reported the production of a monoclonal antibody against anthrax toxin (unpublished data). The monoclonal antibody was shown to have 50-fold improved affinity for anthrax toxin compared with the original fragments and was better than most natural antibodies. It also protected rats against a lethal challenge with the toxin.

A mutant PA molecule has been shown to form part of the doughnut-like normal PA but it was not able to disrupt the membrane pore, thus preventing escape of EF and LF [29]. Rats injected with LF and mutant PA survived. On the basis of these findings a drug based on mutant PA has the potential use in the treatment of anthrax. Another approach studied was the design of a polyvalent inhibitor of anthrax toxin [65]. A peptide isolated from a phage-display library binds weakly to the heptameric cell-binding subunit of anthrax toxin. This prevents interaction between cell-binding and enzymatic moieties. A polyvalent molecule of this non-natural peptide, covalently linked to a flexible backbone, prevented assembly of the toxin complex *in vitro* and blocked the effects of the toxin in an animal model. Recent characterization of the crystal structure of lethal factor and edema factor and the cellular receptor for protective antigen [23–25] will aid identification of drugs that interfere with the binding and the activity of anthrax toxin. Schuch et al. sought bacterioophage lysins able to detect and kill *B. anthracis* [66]. The phage lysin γ (PlyG lysin) isolated from the γ phage of *B. anthracis* was shown to specifically kill *B. anthracis* and other members of the *B. anthracis* "cluster" *in vitro* and *in vivo*. Vegetative cells and germinating spores were susceptible to the lysin. The lytic specificity of PlyG was shown to rapidly identify *B. anthracis*.

5.5.9
Human Vaccination

The current human anthrax vaccine in the UK and the US consists of alum-precipitated cell-free filtrate of an avirulent noncapsulated strain that produces protective antigen [14, 67]. The US vaccine is adsorbed on aluminum hydroxide (also called adsorbed anthrax vaccine or AVA). The vaccine, licensed in 1970, is currently manufactured by the BioPort Corporation (Lansing, Michigan, USA) and given in a series of six inoculations over 18 months. Several animal studies have shown the efficacy of pre-exposure vaccination with AVA [68, 69]. In 1950 a predecessor vaccine to AVA was shown to be 92.5 % efficacious against human cutaneous anthrax in a placebo-controlled trial [70]. The efficacy of AVA after inhalation has been studied in monkeys [58]. After exposure to $8 \times LD_{50}$ of *B. anthracis* spores, 9 of 10 control animals and 8 of 10 animals treated with vaccine alone died. All nine animals receiving doxycycline for 30 days plus vaccine at baseline and day 14 after exposure, survived even after being rechallenged. The US Department of Defense initiated the compulsory anthrax vaccine immunization program to immunize 2.4 million military personnel in 1997 [67]. A report from the Institute of Medicine (IOM) concluded that AVA is effective against inhalational anthrax and if given with appropriate antibiotic therapy may prevent development of the disease after exposure [71]. The report concluded that AVA is acceptably safe. Although contamination with *Mycoplasma fermentans* was suggested as a possible reason for human illness specifically associated with Persian Gulf syndrome, the possibility of mycoplasma contamination of AVA administered to military personnel was subsequently discounted [72]. Testing of vaccine samples by nonmilitary laboratories was negative for viable mycoplasma and mycoplasma DNA and did not support its survival.

The pre-exposure civilian use of AVA in the US is currently limited to people at high risk of exposure to contaminated materials or environments. These include laboratory personnel working with environmental specimens and performing confirmatory testing for *B. anthracis* in the US Laboratory Response Network, workers making repeated entries to known spore-contaminated areas, and other workers who may be repeatedly exposed to aerosolized spores [2]. For several reasons, including unavailability of the vaccine, AVA was not initiated immediately for persons with potential exposure to *B. anthracis* during the 2001 anthrax attacks in the US. It was given later under investigational new drug procedures as an adjunct to the 60 day post-exposure antibiotic prophylaxis [14]. The US Department of Health and Human Services (DHHS) provided three preventive options for persons at risk of inhalation anthrax [73]:

1. Sixty days of antimicrobial prophylaxis accompanied by monitoring for illness and adverse events;
2. Forty additional days of antimicrobial prophylaxis (intended to provide protection against the possibility that anthrax spores may cause illness up to 100 days after exposure) accompanied by monitoring for illness and adverse reactions; and
3. Forty additional days of antimicrobial prophylaxis plus 3 doses of AVA administered over a 4 week period. Because the vaccine is not approved by the FDA for post-exposure prophylaxis, it should be administered with informed consent.

5.5.10
Anthrax Vaccines in Development

Possible future anthrax vaccines are being studied with the intention of making vaccine stocks more plentiful, producing vaccines that require fewer inoculations, and development of a rapid protective response. A vaccine based on recombinant PA produced by non-spore forming *B. anthracis* protects rhesus monkeys against inhalational anthrax [74]. Results also suggests that fewer injections might be needed to elicit an effective immune response and the vaccine might have fewer side-effects than AVA. The recombinant anthrax vaccine is being developed under a fast-track program. Avant Immunotherapeutics based in Needham, MA, USA, is developing an oral one-dose anthrax vaccine. The vaccine is made from PA-producing attenuated *Vibrio cholerae* and acts rapidly.

Although a critical level of vaccine-induced IgG antibodies against the protective antigen is known to confer immunity to anthrax, the role of IgG antibody against the poly γ-d-glutamic acid (γ_DPGA) capsule in protective immunity is not known. Schneerson et al. [75] recently used the nonimmunogenic γ_DPGA or corresponding synthetic peptides bound to BSA, recombinant *B. anthracis* PA (*r*PA), or recombinant *Pseudomonas aeruginosa* exotoxin A (*r*EPA) as immunogens. The anti-γ_DPGA antibodies induced opsonophagocytic killing of capsule-positive and toxin-negative *B. anthracis*. The γ_DPGA-*r*PA conjugates induced both anti-PA and anti-γ_DPGA antibodies. Such conjugates may expand the immune response and enhance the protection conferred by vaccines based on PA alone. A dually active anthrax vaccine

(DAAV) confers simultaneous protection against the replicating bacilli and the toxin [76]. Conjugation of capsular γ_DPGA to PA has been shown to convert the weakly immunogenic PGA to a potent immunogen and to synergistically enhance the humeral response to PA. Complement-mediated killing of the encapsulated bacilli was promoted by PGA-specific antibodies. PA-specific antibodies neutralized the activity of anthrax toxin and protected immunized mice against lethal challenge with the toxin. DAAV introduces a novel vaccine design with potentially wide application against infectious diseases including those related to bioterrorism.

5.5.11
Infection Control and Decontamination

Standard barrier precautions are recommended for patients hospitalized with any form of anthrax. In addition, contact isolation precautions should be used for patients with cutaneous anthrax with draining lesions [3]. Dressings removed from the draining lesions should be incinerated, autoclaved, or otherwise disposed of as biohazardous waste. Only healthcare workers, household contacts, or other contacts determined to have been exposed to the aerosol or surface contamination at the time of an attack should receive post-exposure prophylaxis [14]. When anthrax is suspected, the hospital epidemiologist, the hospital microbiology laboratory, and the state health department should be notified. Safe handling of specimens in the laboratory requires biosafety level 2 conditions and referral to the nearest facility in the Laboratory Response Network under appropriate handling and shipping conditions. Persons coming into direct physical contact with a substance alleged to contain *B. anthracis* should thoroughly wash the exposed skin and clothing with soap and water. For cleaning of environmental surfaces contaminated with infected body fluids, a disinfectant such as hypochlorite, used for standard hospital infection control, is adequate. Human and animal remains should be handled appropriately to prevent further transmission of the disease. Special risks should be considered during embalming of the bodies. Cremation should be considered in preference to burial [14]. All autopsy-related instruments and materials should be autoclaved or incinerated.

Decontamination is usually defined as a procedure causing irreversible inactivation of infectious agents so that a contaminated article or area is rendered safe. Bacterial spores, for example those of *B. anthracis*, may not be eliminated by conventional decontamination procedures. The sporicidal activity of different agents may be affected by variations in time, temperature, concentration, pH, and relative humidity [77]. The US Environmental Protection Agency (EPA) recommends the use of sodium hypochlorite as a sporicidal agent under an emergency exemption because it should be used under specified conditions [78]. The sporicidal effectiveness of hypochlorite solution depends on the concentration of free available chlorine and on pH. The pH of common household bleach (sodium hypochlorite) is 12, to prolong its shelf life. For effective sporicidal activity, bleach must be diluted with water to increase the free available chlorine and with acetic acid to change the pH of the solution to 7. The sporicidal activity of sodium hypochlorite may be reduced by organic matter. Formaldehyde solution or gas has been used both for disinfection and for chemical sterilization. Fumigation with formaldehyde vapor has been used to decontaminate a textile mill. Contamination with *B. anthracis* spores was greatly reduced immediately after treatment and was undetectable after 6 months. Contamination of Gruinard Island, Scotland, with spores occurred during British military testing of explosives in World War II. Spores persisted and remained viable for 36 years after the experiments. Decontamination of the island was conducted in stages between 1979 and 1987, when it was declared fully decontaminated. Materials used during this lengthy process included 280 tons formaldehyde and 2000 tons of sea water. The carcinogenic properties of formaldehyde can be reduced by neutralization with ammonium bicarbonate after fumigation. Gamma radiation was used in the 1960s and 1970s to disinfect baled goat hair contaminated with *B. anthracis* [77]. On the basis of a study by Horne et al., two megarads of gamma radiation are recommended to kill most resistant spores and include a margin of safety [79]. This method was used to decontaminate all mail from contaminated US Postal facilities in 2001.

The greatest risk to humans from an aerosol of *B. anthracis* spores occurs from primary aerosolization, when spores first are made airborne. There is evidence to suggest that after outdoor aerosol release, the threat from exposure would be similar for persons indoor and outdoors [13]. Kournikakis et al. demonstrated that even "low-tech" delivery systems such as opening of envelopes containing powdered spores in an indoor environment can rapidly deliver high concentrations of spores to people in the vicinity [80]. During the 2001 attacks illness was discovered even in persons who handled or processed unopened letters in Washington, D. C. [14]. These cases showed that *B. anthracis* spores of "weapons grade" quality would be capable of leaking from the edges or pores of envelopes. The two fatal inhalational anthrax cases in New York and Connecticut during the 2001 attacks were speculated to have been caused by inhalation of small numbers of spores present in cross-contaminated mail. The risk from secondary aerosolization (resuspension of spores in the atmosphere) is uncertain and depends on many variables. Although the question of illnesses from secondary aerosolization in Sverdlovsk has been debated, the epidemic curve is typical of that for a common-source epidemic with virtually all confirmed cases having occurred within the area of the plume on the day of the accident. The risk of secondary aerosolization was assessed by the US Environmental Protection Agency in the office of Senator Daschle in the Hart Building in Washington, D. C. The experiments demonstrated that routine activity in an environment contaminated with *B. anthracis* spores could cause significant spore suspension [81]. Although these findings do not enable conclusions to be reached about the specific risk of occupants developing anthrax infection in this context, they do have important implications for decontamination, respiratory protection, and reuse of contaminated buildings.

To decontaminate buildings and their contents, multiple technology may be needed. Although the methods used to decontaminate or sterilize laboratories or food industry settings can be used to decontaminate buildings, they have not been scientifically tested in this setting. Decontamination of sections of the Hart Senate Office Building in Washington, D. C. after the opening of a letter laden with *B. anthracis* has been reported to have cost $23 million.

Further research is needed to better characterize risks posed by environmental contamination of spores, particularly inside buildings, and to evaluate methods for environment cleaning after a release.

References

1 Cleri, D. J., Vernaleo, J. R., Rabbat, M. S., et al. **1995** Anthrax. *Infectious Disease Practice*. 77–79.

2 Lucey, D. **2005** *Bacillus anthracis* (Anthrax) In Mandell, Douglas, and Bennett's Principles and Practice of Infectious Diseases. Churchill, Livingstone, Philadelphia, Pennsylvania 2485–2491.

3 Lew, D. **2000** *Bacillus anthracis* (Anthrax) In Mandell, Douglas, and Bennett's Principles and Practice of Infectious Diseases, Churchill, Livingstone, Philadelphia, Pennsylvania 2215–2220.

4 Exodus 9:1–12.

5 Turnbull, P. C. **1991** Anthrax vaccines: past, present and future. *Vaccine*; 9:533–539.

6 Davies, J. C. **1982** A major epidemic of anthrax in Zimbabwe. *Centr Afr J Med*; 18:291–298.

7 Myenye, K., Siziya, S. and Peterson, D. **1996**. Factors associated with human anthrax outbreak in the Chikupo and Ngandu villages of Murewa district in Mashonaland East Province. *Centr Afr J Med*; 42:312–315.

8 Cieslak, T. J. and Eitzen, E. M. **1999** Clinical and epidemiological principles of anthrax. *Emerg Inf Dis* 5:552–555.

9 Turnbull, P. C. **1999** Guidelines for the surveillance and control of anthrax in human and animals. World Health Organization: Emerging and other communicable diseases, surveillance and control. http//www.who.int/csr/resources/publications/anthrax/WHO_EMC_ZDI-98–6/en/

10 Coker, P. R., Smith, K. L. and High-Jones, M. E. **1998** Anthrax in the USA. Proceedings of the Third International Conference on Anthrax, Plymouth, England, September, 7–10; 44 (abstract).

11 Titball, R. W., Turnbull, P. C. and Hutson, R. A. **1991** The monitoring and detection of *B. anthracis* in the environment. *J Appl Bacteriol Symposium* Suppl. 70:9S– 18S.

12 WHO World anthrax data site. www.vetmed.Isu.edu/whocc/mp_world.htm.

13 Inglesby, T. V., Henderson, D. A. Bartlett, J. G., et al. **1999** Anthrax as a biological weapon. Medical and Public health management. *JAMA* 281:1735–1745.

14 Inglesby, T. B., O'Toole, T., Henderson, D. A., et al. **2002** Anthrax as a biological weapon, 2002. Updated recommendations for management, *JAMA* 287:2236–2252.

15 CDC **2003** Laboratory response network (LRN). Level A laboratory procedures for identification of *Bacillus anthracis*. P 1–18. www.bt.cdc.gov/agent/anthrax/LevelAProtocol./anthraxlabprotocol.pdf

16 Read, T. D., Salzberg, S. L., Pop, M., et al. **2002** Comparative genome sequencing for discovery of novel polymorphism in Bacillus anthracis. *Science* 296:2028–2033.

17 Farrar, W. E. **1994** Anthrax: virulence and vaccines. *Ann of Int Med* 121:379–380.

18 Green, B. D., Battisti, L. Koeler, T. M., et al. **1985** Demonstration of a capsule plasmid in *Bacillus anthracis*. *Infect Immun* 49:291–297.

19 Mikesell, P., Ivins, B. E., Ristroph, J. D., et al. **1983** Evidence for plasmid-mediated toxin production in *Bacillus anthracis*. *Infect Immun* 39:371–376.

20 Beauregard, K. E., Collier, R. J. and Swanson, J. A. **2000** Proteolytic activation of receptor-bound anthrax antigen on macrophages promotes its internalization. *Cellular Microbiology* 2:251–258

21 Okinaka, R. T., Cloud, K. Hampton, O., et al. **1999** Sequence and organization of pX01, the large *Bacillus anthracis* plasmid harboring the anthrax toxin genes. *J of Bacteri* 181:6509–6515.

22 Okinaka, R. T., Cloud, K., Hampton, O., et al.**1999** Sequence, assembly and analysis of pX01 and pX02. *J of App Microb* 87:261–262.

23 Bradley, K. A., Mogridge, J. Mourez, M., et al. **2001** Identification of the cellular receptor for anthrax toxin. *Nature* 414:225–229.

24 Drum, C. L., Yan. S.Z, Bard, J., et al. **2002** Structural basis for the activation of anthrax adenylyl cyclase exotoxin by calmodulin. *Nature* 415:396–402.

25 Pannifer, A. D., Wong, T. Y. Schwarzenbacher, R., et al. **2001** Crystal structure of the anthrax lethal factor. *Nature* 414:229–230.

26 Agrawal, A. Lingappa, J., Leppla, S. H., et al.**2003** Impairment of dendritic cells and adaptive immunity by anthrax lethal toxin. *Nature* 424:329–333.

27 Erwin J., DaSilva, L. M., Bavar, S. Little, S. F., et al. **2001** Macrophage-derived cell lines do not express proinflammatory cytokines after exposure to *Bacillus anthracis* lethal toxin. *Inf and Immun* 2:1175–1177.

28 Young, J. A. and Collier, R. J. **2002** Attacking anthrax. *Sci Am* (March) 48–59.

29 Sellman, B. R., Mourez, M., Collier, R. J. **2001** This time it was real – Knowledge of anthrax put to the test. *Science* 292:695–697.

30 Hanna, P. C., Acosta, D. and Collier, R. J. **1993** On the role of macrophages in anthrax. *Proc. Natl. Acad. Sci. USA.* 90:10198–10201.

31 Moayeri, M., Haines, D. Young, H.A., et al. **2003** *Bacillus anthracis* lethal toxin induces TNF-α|-independent hypoxia-mediated toxicity in mice. *J Clin Invest* 112:670–682.

32 Prince, A. S. **2003** The host response to anthrax lethal toxin: Unexpected observations. The *J of Clin Invest* 112:656–658.

33 Brachman, P., and Friedlander, A, **1999** Anthrax: In *Vaccines*, Plotking S. Orenstein, W. eds WB Saunders Co. Philadelphia, Pennsylvania 629–637.

34 Centers for Disease Control and Prevention. **1994** Summary of notifiable diseases, 1945–1994 *MMWR* 43:70–78.

35 American Academy of Dermatology. Anthrax www.aad.org/BioInfo/anthrax.html

36 The Universidad Peruana Cayetano Heredia Gorgas Course in Clinical Tropical Medicine http//info.dom.edu/gorgas/anthrax/html

37 Kanafani, Z. A., Ghosssain, A. Sharara, A. I. et al. **2003** Endemic gastrointestinal anthrax in 1960's Lebanon: Clinical manifestations and surgical findings. *Emerg Infect Dis* 9:520–525.

38 Centers for Disease Control and Prevention **2000** Human ingestion of *Bacillus anthracis*-contaminated meat – Minnesota. *MMWR* 49:813–816.

39 Borio, L. Frank, D. Mani, V. et al. **2001** Death due to bioterrorism-related inhalational anthrax. Report of 2 patients. *JAMA* 286:2554–2559.

40 Lincoln, R. Hodges, D. Klein, F. et al. **1965** Role of the lymphatics in the pathogenesis of anthrax. *J Infect Dis* 115:481–494.

41 Sirisanthana, T., Navachareon, N., Tharavichitkul, P., et al. **1984**. An Outbreak of oral–pharyngeal anthrax. *Am J Trop Med Hyg* 33:144–150.

42 Abramova, F. A., Grinberg, L. M., Yampooskaya, O., et al. **1993** Pathology of inhalational anthrax in 42 cases from the Sverdlovsk outbreak in 1979. *Proc Natl Acad Sci USA* 90:2291–2294.

43 Brachman, P. and Friedlander, A. **1980** Inhalation anthrax. Ann *NY Acad Sci* 353:83–93.

44 Brookmeyer, R. Blades, N. Hugh-Jones, M. and Henderson, D. **2001** The statistical analysis of truncated data: application to the Sverdlovsk anthrax outbreak. *Biostatistics* 2:233–247.

45 Henderson, D. W. Peacock, S. and Belton F. C. **1956** Observations on the prophylaxis of experimental pulmonary anthrax in the monkey. *J Hyg* 54:28–36.

46 Kyriacou, D. K., Stein, A. C., Yarnold, P. R., et al. **2004** Clinical predictors of bioterrorism-related inhalational anthrax. *Lancet* 364:449–452.

47 Peters, C. J. and Hartley, D. M. **2002** Anthrax inhalation and lethal human infection. *The* Lancet 359:710.

48 Dahlgren, C. M., Buchanan, L. E. Decker, H. M., et al. **1960** *Bacillus anthracis* aerosols in goat hair processing mills. *Am J Hyg* 72:24–31.

49 Meyer, M. A. **2003** Neurological complications of anthrax: A review of the literature. *Archives of Neurology* 60:483–488.

50 Lanska, D. J. **2002** Anthrax meningoencephalitis. *Neurology* 59:327–334.

51 Bush, L. M., Abrams, B. H., Beall, A. and Johnson, C. C. **2001** Index case of fatal inhalational anthrax due to bioterrorism in the United States. *N Engl J Med* 345:1607–1610.

52 Centers for Disease Control and Prevention **2002** http://www.bt.cdc.gov/LabIssues/index.asp

53 Uhl, J. R., Bell, C. A., Sloan, L. M., et al.**2002** Application of rapid-cycle real-time polymerase chain reaction for the detection of microbial pathogens: The Mayo– Roche rapid anthrax test. *Mayo Clinic Proc* 77:673–680.

54 Espy, M. J., Uhl, J. R., Sloan, L. M., et al. **2002** Detection of Vaccinia virus, herpes simplex virus, Varicella-zoster virus, and *Bacillus anthracis* DNA by LightCycler polymerase chain reaction after autoclaving: Implications for biosafety of bioterrorism agents. *Mayo Clinic Proc* 77:624–628.

55 Dixon, T. C., Messelson, M., Guillemin, J., et al. **1999** Anthrax *N Engl J Med* 341:815–826.

56 Shlyakhov, E. and Rubenstein, E. **1996** Evaluation of the anthraxin skin test for diagnosis of acute and past human anthrax. *Eur J Clin Microbiol Infect Dis* 15:242–245.

57 Franz, D. R., Jahrling, P. B., Friedlander, A. et al. **1997** Clinical recognition and management of patients exposed to biological warfare agents. *JAMA* 278:399–411.

58 Friedlander, A. M., Welkos, S. L., Pitt, M. L. et al. **1993** Post-exposure prophylaxis against experimental inhalation anthrax. *J Infect Dis.* 167:1239–1242.

59 American Hospital Formulary Service **1996** AHFS Drug Information. Bethesda, MD American Society of Health System Pharmacists. Bethesda, Maryland.

60 Choe, C., Bouhaouala, S. Brook, I., et al. **2000** *In vitro* development of resistance to Ofloxacin and doxycycline in *Bacillus anthracis* Sterne. *Antimicrob Agents Chemother* 44:1766.

61 Jernigan, J., Stephens, D., Ashford, D., et al. **2001** Bioterrorism–related inhalation anthrax: the first 10 cases reported in the United States. *Emerg Infect Dis* 7:933–944.

62 Stevens, D. L., Gibbons, A. E., Bergstron, R., et al. **1988** The Eagle effect revisited. *J Infect Dis* 158:23–28.

63 Enserink, M. **2002** Borrowed Immunity' may save future victims. *Science* 295:777.

64 Maynard, J. A., Maassan, C. B., Leppla, S. H., et al. **2002** Protection against anthrax toxin by recombinant antibody fragments correlates with antigen affinity. *Nature Biotechnol* 20:597–601.

65 Mourez, M., Kane, R. S., Mogridge, J., et al.**2001** Designing a polyvalent inhibitor of anthrax toxin. *Nature Biotechnology* 19:958–961.

66 Schuch, R., Nelson, D. and Fischetti, V.A. **2002** A bacteriolytic agent that detects and kills *B. anthracis*. *Nature* 418:884–888.

67 Jefferson, T. **2004** Bioterrorism and compulsory vaccination. *BMJ* 329:524–525.

68 Ivins, B. E., Fellows, P., Pitt, M. L. et al. **1996** Efficacy of standardized human anthrax vaccine against *Bacillus anthracis* aerosol spore challenge in rhesus monkeys. *Salisbury Med Bull.* 87:125–126.

69 Fellows, P., Linscott, M., Ivins, B. et al. **2001** Efficacy of a human anthrax vaccine in guinea pigs, rabbits, and rhesus macaques against challenge by *Bacillus anthracis* isolates of diverse geographical origin. *Vaccine* 20:635.

70 Brachman, P. S., Gold, H., Plotkins, S. A., et al. **1962** Field evaluation of human anthrax vaccine. *Am J Public Health* 52:632–645.

71 Committee to assess the safety and efficacy of the anthrax vaccine, Medical Follow-Up Agency. **2002** The anthrax vaccine: Is it safe? Does it work? Institute of Medicine, National Academy Press, Washington, D. C. Available at: http://www.iom.edu/iom/iomhome.nsf/WFiles/Anthrax-8-pager1FINAL/$file/Anthrax-8-pager1FINAL.pdf

72 Hart, M. K., Del Giudice, R. A. and Korch, G. W., Jr. **2002** Absence of mycoplasma contamination in the anthrax vaccine. *Emerg Inf Dis* 8:94–96.

73 Centers for Disease Prevention and Control **2001** Additional options for preventive treatment for person exposed to inhalational anthrax. *MMWR* 50:11–42.

74 Friedlander **2002** New anthrax vaccine gets a green light. *Science* 296:639–640.

75 Schneerson, R., Kubler-Kielb, J., Liu, T. Y., et al. **2003** Poly (γ-d-glutamic acid) protein conjugates induce IgG antibodies in mice to the capsule of *Bacillus anthracis*: A potential addition to the anthrax vaccine. *Proc Natl Acad Sci* 100:8945–8950.

76 Rhie, G. E., Roehrl, M. H., Mourez, M., et al. **2003** A dually active anthrax vaccine that confers protection against both bacilli and toxins. *Proc. Natl. Acad. Sci* 100:10925–10930.

77 Spotts-Whitney, E. A., Beatty, M. E., Taylor, T. H., et al. **2003** Inactivation of *Bacillus anthracis* spores. *Emerg Inf Dis* 9:623–627.

78 U.S. Environmental Protection Agency. **2003**Pesticides: topical and chemical fact sheets.[cited 2003 March 31]. Available from: URL: http://www.epa.gov/pesticides/factsheets/chemicals/bleachfact-sheet.htm#bkmrk7

79 Horne, T., Turner, G. and Willis A. **1959** Inactivation of spores of *Bacillus anthracis* by γ-radiation. *Nature* 4659:475–476.

80 Kournikakis, B., Armour, S. J., Boulet, C. A., et al. **2001** Risk assessment of anthrax threat letters. Defense Research Establishment Suffield. Available at http://www.dres.dnd.ca/Meetings/FirstResponders/tr01–048an-nex.pdf

81 Weis, C. P., Intrepido, A. J., Miller, A. K., et al. **2002** Secondary aerosoli-zation of viable *Bacillus anthracis* spores in a contaminated US Senate office. *JAMA* 288:2853–2858.

6
Plague: Endemic, Epidemic, and Bioterrorism

Janak Koirala

6.1
Introduction

Plague is a zoonosis caused by *Yersinia pestis*. It maintains its natural cycle between rodents and fleas, which serve as reservoirs and vectors, respectively. Human infection can lead to the bubonic, pneumonic, or septicemic plague, depending on the type of exposure and host factors. Historically it has caused numerous epidemics and three recorded pandemics resulting in the death of millions. Because it can be potentially aerosolized and spread to cause pneumonic plague, which is a rapidly transmissible, severe, and fatal illness, *Yersinia pestis* is regarded as a "category A" agent of bioterrorism.

6.2
History

Plague is an ancient disease which can be traced back to the earliest periods of the recorded history of human civilization. The first recorded epidemic of plague was the outbreak among the Philistines in 1320 BC, which has been described in the Bible. The first known pandemic, Justinian's plague, recorded between 542 and 767 A. D., claimed nearly 100 million lives in Asia, Africa, and Europe. The second known pandemic, called the Black Death, occurred in the fourteenth century (1347–1350) and caused an estimated 50 million deaths, half in Asia and Africa and half in Europe. The third pandemic began in Canton and Hong Kong in 1894 and spread rapidly throughout the world, carried by rats aboard steamships. Within 10 years (1894–1903) plague entered 77 ports on five continents. It caused nearly 13 million deaths in India alone [1–3].

In 1894, the causal agent of plague was discovered, and it was also established that rats contract plague. The rat flea, *Xenopsylla cheopis*, was identified as the common vector.

Bioterrorism Preparedness. Edited by Nancy Khardori
Copyright © 2006 WILEY-VCH Verlag GmbH & Co. KGaA, Weinheim
ISBN: 3-527-31235-8

Phylogenetic analysis suggests that *Yersinia pestis* is a clone that evolved from *Y. pseudotuberculosis* approximately 1,500–20,000 years ago, most probably shortly before the first known pandemics of human plague [4].

6.3
Microbiology

Y. pestis, a member of enterobacteriaceae family, is a nonmotile, nonspore-forming, Gram-negative coccobacillus measuring 1.5×0.75 μm. When stained with aniline dyes, the ends of the bacillus take stain more intensely, giving the appearance of bipolar staining ("closed safety pin"). *Y. pestis* belongs to the group of bacilli with low resistance to environmental factors such as sunlight, high temperatures, and desiccation. Ordinary disinfectants such as lysol and preparations containing chlorine kill it within 1 to 10 min [5].

Y. pestis grows on sheep's blood agar (SBA) forming gray–white, translucent colonies which have a raised, irregular "fried egg" appearance after incubation for 48–72 h. They form small, nonlactose-fermenting colonies on MacConkey (MAC) or eosin methylene blue (EMB) agar. In conventional nutrient-rich broths, for example brain–heart infusion (BHI), *Y. pestis* grows in clumps that are typically described as "flocculant" or "stalactite" in appearance [5].

Three biovars of *Y. pestis* are antiqua, medievalis, and orientalis, which correspond to the three pandemics of plague [4]. Biovar antiqua prevalent in Africa, Southeastern Russia, and Central Asia, is thought to be the cause of the first pandemic (Justinian's plague). Biovar mediavalis, prevalent in Caspian Sea region, is thought to be the cause of Black Death. The third biovar, orientalis, is the cause of the third pandemic (modern plague) and is still circulating in Asia and the Western hemisphere. The complete genomes of *Y. pestis* (strains CO92 and KIM) have been sequenced.

6.4
Global Epidemiology

According to the World Health Organization, between 1954 and 1997, human plague was reported from 38 countries. Over this 44 year period, 80,613 cases of plague with 6,587 deaths were reported to the WHO. The largest proportion of cases (58%) was from Asia and included epidemics in Vietnam and India [6, 7]. In the USA, CDC reported 390 cases over a 50-year period (1947–1996) including 84% bubonic, 13% septicemic, and 2% pneumonic plague cases.

Outbreaks of human plague depend on the maintenance of the disease in an animal reservoir. The natural foci of plague persist in Asia, Africa, North and South America, and to some extent in South-East Europe. In North America, natural foci

World distribution of plague, 1998

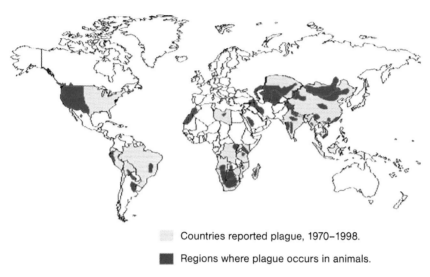

  Countries reported plague, 1970–1998.

  Regions where plague occurs in animals.

Fig. 6.1
Natural foci of plague (Source: CDC, Atlanta, Georgia, USA).

of plague occur in 15 western states of the USA, in south-western Canada, and in northern Mexico. South American natural foci have been recorded in Argentina, Bolivia, Brazil, Ecuador, Peru, and Venezuela (Fig. 6.1).

Y. pestis is maintained in wild rodents, for example rats, squirrels and prairie dogs, as a zoonosis. Rat fleas serve as vectors for transmission of the disease. Human outbreaks occur when domestic rodents become infected with the bacteria. Rodent-to-human transmission can occur by flea bites and, less commonly, by direct contact with or handling of infected materials. Human-to-human transmission can also occur as a result of human flea bites from septicemic patients, direct contact, or respiratory exposure to patients with pneumonic plague.

Human cases of plague are relatively sparse in natural foci. Cases occur among people who come in contact with wild rodents in the course of their work, hunting, or camping. The risk of human infection increases significantly when plague penetrates populations of domestic rats, particularly *Rattus spp.*

6.5
Pathogenesis

A bite from a plague-infected flea deposits thousands of organisms (*Y. pestis*) in the human skin. *Y. pestis* migrates through cutaneous lymphatics to the regional lymph nodes. Antiphagocytic factors, for example F1, V and W antigens, in *Y.*

pestis enable them to resist destruction by phagocytosis. *Y. pestis* rapidly multiplies in the lymph nodes causing their destruction and necrosis. Subsequently, bacteremia, septicemia, and endotoxemia leading to shock can occur. This may result in disseminated intravascular coagulation (DIC) and coma.

Several virulence factors have been identified in *Y. pestis* [8–10]; these are summarized below (Table 6.1). F1 antigen, encoded in *pFra* plasmid, is an antiphagocytic factor which also elicits humoral response. It is used for immunologic diagnostic tests. Plasminogen activator, encoded in pesticin or *Pst* plasmid, facilitates systemic spread by degrading fibrin and other extracellular proteins. Hemin storage system enhances survival in phagocytes and increases uptake by eukaryocytic cells. It is used as a laboratory marker of the pigmentation system of the bacteria. V and W antigens make *Y. pestis* resistant to phagocytosis. Low-calcium-response (*Lcr*) plasmid activates V-antigen under low-calcium conditions. Yops (*Yersinia* outer proteins) inhibit phagocytosis, platelet aggregation, and effective inflammatory response. *Lcr* plasmid also activates Yops under low-calcium conditions. Lipopolysaccharide endotoxin causes classic endotoxic shock. Phospholipase D (PLD) enables the bacilli to survive in the flea gut.

Tab. 6.1
Virulence factors of *Yersinia pestis*.

F1 Antigen:	An antiphagocytic factor, elicits humoral response, used for immunologic diagnostic tests
Plasminogen activator:	Facilitates systemic spread by degrading fibrin and other extracellular proteins.
Hemin storage system:	Enhances survival in phagocytes, increases uptake by eukaryocytic cells, as a laboratory marker of pigmentation.
V and W antigens:	Makes resistant to phagocytosis.
Low-calcium-response (*Lcr*) plasmid:	Activates V-antigen under low-calcium conditions.
Yops (*Yersinia* outer proteins):	Inhibits phagocytosis, platelet aggregation and effective inflammatory response
Lipopolysaccharide endotoxin:	Causes classic endotoxic shock.
Phospholipase D (PLD):	Enables survival in the flea gut

6.6
Clinical Features

Plague presents in three primary clinical forms – bubonic plague, septicemic plague, and primary pneumonic plague. It may also present with several other clinical manifestations which occur less frequently.

6.6.1
Bubonic Plague

This is the most common presentation of plague. Bubonic plague is characterized by regional lymphadenopathy resulting from cutaneous or mucous membrane exposure, for example a flea bite or direct contamination. The incubation period is 2 to 6 days, which is followed by a sudden onset of illness characterized by headache, shaking chills, fever, malaise, and pain in the affected regional lymph nodes. Progression of symptoms is usually rapid with the regional lymphadenitis becoming very tender and painful. A local cutaneous lesion usually occurs at the site of inoculation, although this may not be clinically evident. Occasionally a vesicle, pustule, or ulcer may develop at the site of inoculation. The bacteria spread via the lymphatics to the regional lymph nodes causing inflammation and swelling of lymph nodes, known as buboes. Buboes may occur in any regional lymph node sites such as inguinal, axillary, supraclavicular, cervical, post-auricular, epitrochlear, popliteal, or pharyngeal. It may also spread to the intraabdominal lymph nodes.

6.6.2
Primary Septicemic Plague

This is characterized by overwhelming *Y. pestis* bacteremia, usually after cutaneous exposure, with the apparent absence of primary lymphadenopathy. Primary septicemic plague occurs in all age groups, with the elderly at the greatest risk. The bacteremia and bacterial endotoxins trigger a widespread immunological cascade resulting in the sepsis syndrome, including disseminated intravascular coagulopathy (DIC), multi-organ dysfunction, and adult respiratory distress syndrome (ARDS). Dissemination of *Y. pestis* may lead to complications such as pneumonia, meningitis, hepatic or splenic abscesses, endophthalmitis, or generalized lymphadenopathy.

6.6.3
Primary Pneumonic Plague

Inhalation of droplets containing *Y. pestis* can result in a primary pulmonary infection after a short incubation period (usually 1–3 days). The illness starts with a sudden onset of chills, fever, headache, body aches, weakness, and chest discomfort. Patients develop cough with sputum production, chest pain, shortness of breath, hypoxia, and hemoptysis. It is the most fulminating and fatal form of plague, usually resulting in death within 18–24 h of the onset of illness.

6.6.4
Other Forms

In addition to the three primary forms of plague discussed above, it may present in several other forms including pharyngitis, which results from exposure of the oropharynx with *Y. pestis*-contaminated materials. Asymptomatic colonization of the pharynx has also been reported in contacts of pneumonic plague. Meningitis may also occur as a presentation of primary plague infection.

6.7
Mortality

On the basis of cases reported to the WHO, mean perennial plague mortality for the world during the period 1954–1997 was 7.4%, with a range of 2.4–23.8% [6]. Case-fatality rates during 1947–1996 in the USA were: bubonic 14%, septicemic 22%, and pneumonic 57%. Untreated cases of pneumonic plague result in 100% mortality.

6.8
Laboratory Diagnosis

When plague is suspected, collection of clinical specimens and initiation of appropriate antimicrobial therapy should be performed without delay. Routine diagnostic specimens for smear and culture may include blood, aspirates from suspected buboes, and pharyngeal swabs, sputum samples, or tracheal washes from those with suspected plague pharyngitis or pneumonia. Cerebrospinal fluid (CSF) should be obtained from those with suspected meningitis. Microbiology laboratories should be notified about the possibility of plague in suspected cases, because they require Biological Safety Level-2 (BSL-2) practices to be followed [11]. A flow chart suggested for laboratory diagnosis is shown in Fig. 6.2.

Y. pestis appears as plump, Gram-negative coccobacillus, 1–2 μm × 0.5 μm, mostly as single cells, or in pairs and short chains in liquid media. Use of Wright–Giemsa or Wayson stain results in a characteristic bipolar appearance. Direct fluorescent antibody (DFA) testing detects F1 antigen in tissues or fluids, providing presumptive evidence of plague.

Y. pestis grows in a suitable culture media, e.g. brain–heart infusion broth, sheep-blood agar, or MacConkey agar. On solid media it grows as gray–white, translucent colonies, usually too small to be seen at 24 h. After 48–72 hours of incubation colonies are raised and have an irregular, "hammered copper" appearance. Definite identification of *Y. pestis* in culture media can be performed with specific phage lysis. Automated bacteriological test systems can be used to assist with the iden-

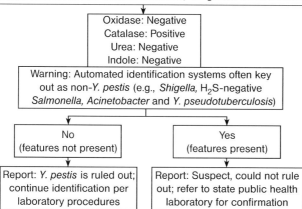

Fig. 6.2
Flow chart for laboratory diagnosis of plague. (Adapted from:
CDC, ASM, APHL (2001) Basic Protocols for Level A Laboratories
for the Presumptive Identification of *Yersinia pestis* [5].)

tification of isolates as *Y. pestis,* but such isolates can be misidentified or overlooked
if these systems are not properly programmed. *Y. pestis* has been falsely identified
as *Y. pseudotuberculosis, Shigella, Salmonella, Acinetobacter,* etc. [5].

A serological diagnosis may be helpful when *Y. pestis* is not isolated from
cultures. By passive haemagglutination testing plague can be confirmed by a
seroconversion with a fourfold or greater change in titer between acute and
convalescent phases to the *Y. pestis* F1 antigen. A single titer of >1:128 is also
considered confirmatory. A single titer of >1:10 should be considered presumptive
positive in a person previously unexposed to the infection or the vaccine [12]. F1
antigen-capture ELISA with 100% sensitivity has been described for bubo specimens; sensitivity for serum and urine is only 52% and 58%, respectively, however [13].

A rapid diagnostic test (RDT) based on monoclonal antibodies against F1 was
tested in Madagascar by a WHO/Pasteur Institute collaborative group. It was
shown to be sensitive, and the specificity was 100%, with positive and negative
predictive values of 91% and 87%, respectively. RDT detected 41% and 31% more
clinical specimens than bacteriological and F1 ELISA methods, respectively [14].
PCR is available only at reference laboratories.

6.9
Radiology

Patchy bronchopneumonic infiltrates and segmental or lobar consolidation, with or without confluence, are usually observed for patients with pneumonic plague. Occasionally, patients may have cavitation, or bilateral diffuse infiltrates consistent with ARDS.

6.10
Potential as a Biological Weapon

Plague has high potential for use as a biological weapon. *Y. pestis* has wide global distribution, and easy mass production and aerosolized dissemination could result in pneumonic plague with a high fatality rate and potential for secondary spread of cases from person to person during an epidemic.

Historically, plague was used to cause outbreaks in enemies as early as the 14th century. In 1346 A. D. the Tartar army hurled its plague-infected corpses over the walls of the city during the siege of Caffa and forced the Genoese defenders to flee. Similar tactics were used by others, including Russian army in the war against Sweden in the 18th century. A secret branch of the Japanese army was reported to have developed and dropped *Y. pestis*-infected fleas and grain over populated areas of China on several occasions during World War II. The grain was used to attract the rats causing outbreaks of plague. Such bombs were used in at least three Chinese cities causing small epidemics of plague [15]. The USA and the Soviet Union subsequently developed techniques to aerosolize plague [11].

In 1970 the World Health Organization (WHO) estimated that if 50 kg *Y. pestis* was released as an aerosol over a city of population five million, pneumonic plague could occur in as many as 150,000 persons, with 36,000 deaths. The plague bacilli would remain viable as an aerosol for one hour for a distance of up to 10 km. Significant numbers of city inhabitants might attempt to flee, further spreading the disease [11].

6.11
Features of Bioterrorism

If used as a weapon of bioterrorism, plague is most likely to occur as an aerosolized attack. Symptoms of pulmonary infection would begin 1 to 6 days after exposure with people dying quickly after the onset of symptoms. Symptoms, for example fever with cough and dyspnea, can easily be confused with those of other severe respiratory illnesses. Gastrointestinal symptoms such as nausea, vomiting, abdominal pain, and diarrhea may also be present.

Bioterrorism should be suspected if plague occurs in places not known to have enzootic infection or in the absence of previous rodent deaths in the area. Foul play should also be suspected if plague is diagnosed in a person without risk factors for plague, for example contact with rodents or fleas, or in multiple individuals without a common source of exposure.

With the development of newer molecular epidemiologic techniques, for example multiple-locus variable-number of tandem repeat analysis (MLVA), when combined with epidemiological information, it would be possible to differentiate naturally occurring cases from those occurring as a result of intentional *Y. pestis* release [16].

6.12
Diagnosis

The US Center for Disease Control and Prevention (CDC) has recommended a set of diagnostic criteria for a uniform notification and surveillance of plague [12]. Case definitions of suspect, presumptive and confirmed cases of plague are as follows (Table 6.2):

1. **Suspect Plague**: Clinical symptoms, for example fever and lymphadenopathy, which are compatible with plague, in a person who resides in or has recently traveled to a plague-endemic area should raise suspicion of plague. If small Gram-negative or bipolar-staining coccobacilli are seen on a smear taken from affected tissues, e.g. a bubo, blood, tracheal or lung aspirate, the microbiology laboratory should consider the specimen as a suspect case of plague and take appropriate steps before further processing of the specimen.

2. **Presumptive Plague**: A presumptive diagnosis of plague should be made if the immunofluorescence stain of smear or material is positive for the presence of *Y. pestis* F1 antigen. A presumptive diagnosis is also made if only a single serum specimen is tested and the *Y. pestis*-specific anti-F1 antigen titer by hemagglutination inhibition is >1:10.

3. **Confirmed Plague**: A case of plague is confirmed if a culture isolate is lysed by *Y. pestis* specific bacteriophage. Alternatively, plague is also serologically confirmed if a fourfold or greater rise in *Y. pestis* specific anti-F1 antigen titer in hemagglutination inhibition test is observed for paired serum specimens taken during acute and convalescent phases. A single serum specimen tested by the plague-specific hemagglutination test with a titer of >1:128, in a patient with no known previous plague exposure or vaccination history, is also considered as a confirmed case of plague.

Tab. 6.2
Diagnostic criteria for plague. (Modified from Center for Disease
Control and Prevention (CDC) case definitions of suspect,
presumptive and confirmed cases of plague, CDC, Atlanta [12].)

---

1. Suspect Plague:

Clinical symptoms compatible with plague (fever and lymphadenopathy) in a person who
resides in or has recently traveled to a plague-endemic area

Small Gram-negative and/or bipolar-staining coccobacilli on a smear from affected tissues, e.g. a
bubo, blood, tracheal or lung aspirate

2. Presumptive Plague:

Immunofluorescence stain (DFA) positive for *Y. pestis* F1 antigen

A single serum specimen with anti-F1 antigen titer >1:10 by agglutination

3. Confirmed Plague:

A culture isolate lysed by specific bacteriophage for *Y. pestis*

Fourfold or greater rise in anti-F1 antigen titer in agglutination test of paired serum specimens
(acute and convalescent)

A single serum specimen tested by agglutination has a titer of >1:128 and the patient has no
known previous plague exposure or vaccination history (The agglutination test must be shown to
be specific to *Y. pestis* F1 antigen by hemagglutination inhibition.)

6.13
Treatment

When a patient is suspected of having plague, appropriate clinical specimens for
diagnosis should be obtained and the patient should be started on specific antimicrobial therapy without delay. Suspect plague patients with evidence of pneumonia should be placed in isolation, and managed under respiratory droplet precautions. It is important to ensure appropriate supportive care. Patients may present
with septicemia with or without septic shock, requiring immediate resuscitation,
hemodynamic monitoring, and fluid-electrolyte balance. Patients may need intravenous fluid resuscitation, vasopressors, intensive care admission, and respiratory
care including ventilator support.

As shown in Table 6.3, streptomycin is the preferred drug for treatment of plague.
It is given at a dose of 30 mg kg^{-1} body weight, administered as an intramuscular
injection, every twelve hours for total of ten days, or until three days after the
temperature returns to normal. Gentamicin has been used as an alternative aminoglycoside. Oral alternatives include tetracyclines and chloramphenicol. A retrospective data analysis of 75 cases from New Mexico showed similar outcomes in patients
who received streptomycin alone, gentamicin alone, or gentamicin in combination
with tetracyclines [17]. Although there are no clinical studies of humans, animal and
in-vitro studies suggest equivalent or higher efficacy of ciprofloxacin compared with

the aminoglycosides [11, 18, 19]. All fluoroquinolones, including ciprofloxacin, ofloxacin, moxifloxacin, and gatifloxacin, which were tested using animal models (mice), were found to have similar therapeutic and prophylactic efficacy [20–22]. On the basis of animal and *in vitro* data the Working Group on Civilian Biodefense has recommended ciprofloxacin as an alternative agent [11].

Tab. 6.3
Recommended antibiotic treatment for plague.

Adults:

Preferred agent:	Streptomycin	30 mg kg^{-1} day^{-1} (up to 1 g twice daily), intra-muscular, for 10 days or until 3 days after the temperature has returned to normal
Alternative agents:	Gentamicin	5 mg kg^{-1} IM or IV once daily, or 2 mg kg^{-1} loading dose then 1.7 mg kg^{-1} IM or IV, three times daily for 10 days
	Tetracyclines	(doxycycline 100 mg IV twice daily or 200 mg IV once daily for 10 days)
	Chloramphenicol	25 mg kg^{-1} IV four times daily for 10 days
	Ciprofloxacin	400 mg IV or 500 mg oral twice daily for 10 days (There are no human studies of the use of fluoroquinolones in plague, but *in vitro* and animal studies suggest efficacy of ciprofloxacin is equivalent to or higher than that of aminoglycosides)

Children:

Preferred agents:	Streptomycin	15 mg kg^{-1} IM twice daily (maximum 2 g day^{-1})
	Gentamicin	2.5 mg kg^{-1} IM or IV three times daily for 10 days
Alternative agents:	Doxycycline	for less than 45 kg, 2.2 mg kg^{-1} IV twice daily for 10 days (maximum 200 mg day^{-1})
	Ciprofloxacin	15 mg kg^{-1} IV twice daily for 10 days (maximum 1 g day^{-1})
	Chloramphenicol	25 mg kg^{-1} IV four times daily for 10 days (maximum 4 g day^{-1})

Pregnancy:	Gentamicin	same as adult dose above

Although sulfonamides have been used extensively for treatment and prevention of plague, some studies have revealed higher mortality, increased complications, and longer duration of fever compared with other agents. Similarly, rifampin, aztreonam, and beta-lactam antibiotics including cephalosporins should not be used to treat plague, because they are ineffective. When the beta-lactam antibiotics were used to treat plague, higher mortality in humans and accelerated death in mice were observed [17, 19]. Plasmid-mediated, multidrug resistant *Y. pestis* has been reported in Madagascar [23]. A few Russian publications have reported ciprofloxacin resistance in virulent laboratory isolates of *Y. pestis* [24, 25].

In children, streptomycin or gentamicin can be used to treat plague. Gentamicin is the preferred antibiotic for treating plague in pregnancy because of its safety, ease of administration (intravenous or intramuscular), and the ability to monitor blood concentrations in most laboratories (Table 6.3).

Oral antibiotics are recommended for a large outbreak or for mass casualties, because most patients do not need parenteral antibiotics (Table 6.4). In such instances, the local healthcare system's ability to handle parenteral antibiotics could be exhausted. The Group on Civilian Biodefense recommends oral doxycycline or ciprofloxacin for children and adults for such a large-scale outbreak. Alternatively, chloramphenicol can be used. Recommended duration of treatment for all three agents is 10 days [11].

Tab. 6.4
Treatment/prophylaxis of plague in the event of mass casualties.

Duration:	Treatment – 10 days		
	Post-exposure prophylaxis – 7 days		
Preferred agents:	Doxycycline:	Adults	100 mg, PO, twice daily
		Children (<45 kg)	2.2 mg kg^{-1}, PO, twice daily
	Ciprofloxacin:	Adults	500 mg, PO, twice daily
		Children	20 mg kg^{-1}, PO, twice daily
Alternative agents:	Chloramphenicol:	Adults	25 mg kg^{-1}, four times daily
		Children	25 mg kg^{-1}, four times daily

6.14
Prevention

Plague is highly infectious with multiple routes of transmission; prevention of transmission is challenging. In natural foci plague exists as a zoonosis and most human outbreaks occur by transmission to domestic rodents and house fleas. Public education for control of rodents using simple measures of environmental sanitation, for example removing sources of rodent food and building rodent-proof houses, can be helpful in reducing transmission to humans. Domestic rodents and fleas around human dwellings can also be controlled with rodenticides and insecticides. Travelers to endemic areas or natural foci should avoid contact with rodents and other possibly infected animals. Protective insect repellents (for example as DEET) should be used while traveling to such places.

In addition to these general control methods, other measures available to prevent transmission of plague, discussed below, include immunization, post-exposure prophylaxis, pre-exposure prophylaxis, and infection-control practices.

6.14.1
Immunization

Live attenuated and formalin-killed vaccines are two types of plague vaccine which have been available for human use since the early 20th century. These vaccines are variably immunogenic and moderately to highly reactogenic. These vaccines have variable efficacy in preventing bubonic plague and do not protect against primary pneumonic plague. They are not useful in outbreaks because it takes approximately a month or more to develop a protective immune response [26, 27]. These two vaccines are currently available in most countries outside the USA for persons who work in close contact with *Y. pestis*, for example laboratory technicians in plague reference and research laboratories and persons studying infected rodent colonies. Manufacture of these plague vaccines was stopped in USA in 1999.

A new vaccine based on two purified recombinant antigens, F1 and V, has been found to be capable of protecting mice against both the bubonic and pneumonic forms of plague. Another vaccine prepared by fusion of F1 and V antigens has been tested in non-human primates by USAMRIID (United States Army Research Institute of Infectious Diseases, MD, USA) and shown to induce a significant level of protection. Both these candidate vaccines were scheduled for human trials in 2005.

6.14.2
Antibiotic Prophylaxis

Individuals exposed to plague should receive antibiotic prophylaxis within six days of exposure. Preferred antimicrobial agents for preventive or abortive therapy are tetracyclines, chloramphenicol, or one of the effective sulfonamides (e.g. sulfamethoxazole) for 7 days.

In the event of an outbreak or mass casualties, for example a community experiencing a pneumonic plague epidemic, the Working Group on Civilian Biodefense recommends that all persons developing a temperature of 38.5 °C or higher, or a new cough, should promptly begin taking a parenteral or an oral antibiotic, in accordance with the mass casualty recommendations (Table 6.4). Preferred oral antibiotics are doxycycline or ciprofloxacin; an alternative is chloramphenicol. Prophylaxis should be given for 7 days [11].

Pre-exposure antibiotic prophylaxis may be indicated for persons who must travel to a plague-active area for a short duration under circumstances in which exposure to plague sources (fleas, pneumonic cases) is inevitable.

6.14.3
Infection Control

Human to human transmission of pneumonic plague occurs via respiratory droplets which can be transmitted to contacts within a two-meter distance. In addition to the standard (universal) precautions, droplet precautions are required. These include respiratory isolation, use of standard surgical masks, and avoidance of close contact. These measures should be followed for the first 48 h from initiation of antibiotics or until clinical improvement.

Before sending any clinical specimens with suspected *Y. pestis* infection, microbiology laboratory personnel should be alerted, because they will require BSL-2 precautions. Procedures potentially resulting in aerosol or droplet production require BSL-3 laboratory conditions. Autopsy procedures likely to cause aerosols, e.g. bone sawing, must be avoided. Exposed persons should wash with soap and plenty of water and post-exposure antibiotic prophylaxis should be administered.

Y. pestis is very sensitive to the action of sunlight and heating. It does not survive very long outside the host. It has been estimated that in a worst case a plague aerosol would be effective and infectious for as long as an hour (WHO). Environmental decontamination can be achieved with 0.5 % hypochlorite solution or a 1:10 dilution of household bleach.

All suspected and confirmed cases of plague should be reported to the hospital epidemiologist or infection control practitioner, and to the local or state health departments. More information about plague and reporting can be found on the CDC website at http://www.cdc.gov/ and WHO website at http://www.who.int/csr.

References

1 Achtman M, Morelli G, Zhu P, et al. **2004** Microevolution and history of the plague bacillus, *Yersinia pestis*. Proc Natl Acad Sci USA 101(51):17837–42.

2 Kostis KP. **1998** In search of the plague. The Greek peninsula faces the black death, 14th to 19th centuries. Dynamis 18:465–78.

3 Dols MW. **1979** The second plague pandemic and its recurrences in the Middle East: 1347–1894. J Econ Soc Hist Orient 22(2):162–89.

4 Achtman M, Zurth K, Morelli G, et al. **1999** *Yersinia pestis*, the cause of plague, is a recently emerged clone of *Yersinia pseudotuberculosis*. Proc Natl Acad Sci USA 96(24):14043–8.

5 CDC, ASM, APHL **2001** Basic Protocols For Level A Laboratories for the Presumptive Identification of *Yersinia pestis*, CDC, Atlanta, Georgia.

6 WHO Report **1999** on Global Surveillance of Epidemic-prone Infectious Diseases, World Health Organization, Geneva. WHO/CDS/CSR/ISR/2000.1.

7 Dennis DT, Gage KL, et al. **1999** Plague Manual: Epidemiology, Distribution, Surveillance and Control, WHO, Geneva. WHO/CDS/CSR/EDC/99.2.

8 McGovern TW, Friedlander A. **1997** Plague. In: Zajtchuk R, Bellamy RF, eds. Medical Aspects of Chemical and Biological Warfare. Office of the Surgeon General, Bethesda, Maryland: 479–502.

9 Dennis D, Meier F. Plague.**1997** In: Horsburgh CR, Nelson AM, eds. Pathology of Emerging Infections. ASM Press, Washington, DC: 21–47.

10 Perry RD, Fetherston JD. **1997** *Yersinia pestis* – etiologic agent of plague. Clin Microbiol Rev 10(1):35–66

11 Inglesby TV, et al. **2000** Plague as a Biological Weapon. JAMA 283:2281–2290.

12 Laboratory Testing Criteria for Diagnosis of Plague. CDC, Atlanta, last revised in 4/2005. Available at: http://www.cdc.gov/ncidod/dvbid/plague/lab-test-criteria.htm. Accessed on 8/7/2005.

13 Chanteau S, Rahalison L, Ratsitorahina M, et al. **2000** Early diagnosis of bubonic plague using F1 antigen capture ELISA assay and rapid immunogold dipstick. Int J Med Microbiol 290(3):279–83.

14 Chanteau S, Rahalison L, Ratsitorahina M, et al. **2003** Development and testing of a rapid diagnostic test for bubonic and pneumonic plague. Lancet 361:211–16.

15 US Army. USAMERIID's Medical Management of Biological Casualties Handbook. US Army Medical Research Institute of Infectious Diseases, Fort Detrick, MD.

16 Lowell JL, Wagner DM, Atshabar B, et al. **2005** Identifying sources of human exposure to plague. J Clin Microbiol 43(2):650–6.

17 Boulanger LL, Ettestad P, Fogarty JD, et al. **2004** Gentamicin and tetracyclines for the treatment of human plague: review of 75 cases in new Mexico, 1985–1999. Clin Infect Dis 38(5):663–9.

18 Bonacorsi SP, Scavizzi MR, Guiyoule A, et al. **1994** Assessment of a fluoroquinolone, three beta-lactams, two aminoglycosides, and a cycline in treatment of murine *Yersinia pestis* infection. Antimicrob Agents Chemother 38(3):481–6.

19 Byrne WR, Welkos SL, Pitt ML, et al. **1998** Antibiotic treatment of experimental pneumonic plague in mice. Antimicrob Agents Chemother Mar; 42(3):675–81.

20 Steward J, Lever MS, Russell P, et al. **2004** Efficacy of the latest fluoroquinolones against experimental *Yersinia pestis*. Int J Antimicrob Agents 24(6):609–12.

21 Russell P, Eley SM, Green M, et al. **1998** Efficacy of doxycycline and ciprofloxacin against experimental *Yersinia pestis* infection. J Antimicrob Chemother 41(2):301–5.

22 Russell P, Eley SM, Bell DL, et al. **1996** Doxycycline or ciprofloxacin prophylaxis and therapy against experimental *Yersinia pestis* infection in mice. J Antimicrob Chemother 37(4):769–74.

23 Galimand M, Guiyoule A, Gerbaud G, et al.**1997** Multidrug resistance in *Yersinia pestis* mediated by a transferable plasmid. N Engl J Med 337(10):677–80.

24 Ryzhko IV, Shcherbaniuk AI, Skalyga EIu, et al. **2003** Formation of virulent antigen-modified mutants (Fra-, Fra-Tox-) of plague bacteria resistant to rifampicin and quinolones. Antibiot Khimioter 48(4):19–23.

25 Kasatkina IV, Shcherbaniuk AI, Makarovskaia LN, et al. **1991** Chromosomal resistance of plague agent to quinolones. Antibiot Khimioter 36(8):35–7.

26 Titball RW, Williamson ED. **2004** *Yersinia pestis* (plague) vaccines. Expert Opin Biol Ther 4(6):965–73.

27 Jefferson T, Demicheli V, Pratt M. **2000** Vaccines for preventing plague. Cochrane Database Syst Rev (2):CD000976.

7
Botulism: Toxicology, Clinical Presentations and Management

Janak Koirala

7.1
Introduction

Botulism is an illness caused by a toxic enzyme, botulinum toxin, which is produced by an anaerobic Gram-positive bacillus, *Clostridium botulinum*. Botulinum toxin (in Latin, *botulus* = sausage), also commonly known as "Sausage Poison", is the most poisonous substance known to humans. It blocks the neuromuscular junction resulting in flaccid paralysis leading to death. There are seven subtypes of botulinum toxin, which are designated by the letters A to G. Botulism occurs worldwide, as a sporadic illness and in outbreaks, resulting in conditions with severe neuromuscular weakness and high rate of mortality. Preparations containing very small amounts of botulinum toxin are increasingly being used for the treatment of a variety of neuromuscular disorders. Because botulism is a globally prevalent disease and has a high potential of being used as a weapon of bioterrorism, it has become an important disease for the healthcare providers and public health workers.

7.2
History

A German district medical officer and romantic poet, Justinus Kerner (1786–1862), published the first accurate and complete descriptions of botulism during 1817–1822. Kerner described botulinum toxin as a "fatty poison" or as "sausage poison". He performed animal and human experiments, including experiments of the effect of the toxin on himself. He even thought about its therapeutic use two centuries before it was actually used in the treatment of neuromuscular disorders. His significant contributions in describing "sausage poison" gave him the nickname "*Wurst*-Kerner" (or "Sausage Kerner" in German) [1, 2]. In 1870, another German scientist Muller further characterized the "sausage poison" and named it "botulus",

Bioterrorism Preparedness. Edited by Nancy Khardori
Copyright © 2006 WILEY-VCH Verlag GmbH & Co. KGaA, Weinheim
ISBN: 3-527-31235-8

which means "sausage" in Latin. Emile Pierre van Ermengem, a Belgian micro-biologist, discovered the bacteria producing the toxin in 1895. As a therapeutic agent, Alan B. Scott (USA) used it for chemodenervation in monkeys in 1973, and in humans in 1980 [3].

7.3
Epidemiology

Spores of *C. botulinum* are found in the soil worldwide. Food-borne botulism occurs as a result of ingesting food prepared without following proper hygienic measures and stored in anaerobic conditions without following appropriate meth-ods of preservation. Contamination of food with the spores of *C. botulinum* usually occurs during preparation. The ideal conditions under which this organism can grow and produce toxin in the food are an anaerobic environment, acidic pH (usually 4.6 to 4.8), a temperature of 35 °C (minimum 10 °C for most strains), and the availability of water with limited solute concentration [4].

Distribution of botulism is worldwide. Sporadic cases, small family outbreaks, or larger outbreaks occur as a result of contaminated food. For example, in 2001, a total of 169 cases of botulinum intoxication were reported to the CDC. Of these, 112 (66 %) were infant cases, 33 cases (20 %) were food-borne, and 23 were cases of wound botulism. As shown in Table 7.1, type A toxin was found more commonly in food-borne (61 %) and wound botulism (96 %), whereas type B toxin was more common in infant botulism (60 %). Type E toxin was found in wound botulism only. Only one case of adult botulism related to intestinal colonization was reported, which was caused by type F toxin [4].

Tab. 7.1
Distribution of the botulinum toxins in the USA by type, and associated illness.

Type of botulism	Number of cases	Type A	Type B	Type E	Type F
Food-borne	33	20 (61 %)	2 (6 %)	10 (30 %)	1 (3 %)
Wound	23	22 (96 %)	1 (4 %)	–	–
Infant	112	45 (40 %)	67 (60 %)	–	–
Adult Intestinal	1	–	–	–	1 (100 %)
All cases	169	87 (51 %)	70 (41 %)	10 (6 %)	2 (1 %)

Source: CDC Botulism Surveillance Report for the year 2001 [4]

There is no worldwide surveillance system for botulism. In the USA the Center for Disease Control and Prevention (CDC) has kept surveillance records and reported all cases, including outbreaks, since 1899. During the period 1899 to 1996 CDC recorded a total of 921 outbreaks of food-borne botulism in the United States with an average of 9–10 outbreaks per year. A total of 2368 cases of food-borne botulism were reported during this period with approximately 2.5 cases per outbreak. Very little change in this trend was observed over the century, with the number of outbreaks per year remaining approximately the same. Of the 444 outbreaks of food-borne botulism since 1950, 37.6% were caused by type A botulinum toxin, 13.7% by type B, 15.1% by type E and 0.7% by type F. Toxin type in the rest (32.9%) were not identified [5]. A distinct geographic pattern of distribution of *C. botulinum* spores in soil has been found to correlate with the type of botulinum toxins implicated in various outbreaks. For example, during 1950–1996 most of the food-borne outbreaks with type A botulinum toxin occurred in the states west of Mississippi river, whereas most of the type B toxin-related outbreaks occurred in the eastern states [5].

The vehicles for food-borne botulism reported in recent outbreaks were related to homemade canned food prepared from vegetables, fish, and marine animals. Some of the examples include homemade salsa, baked potatoes sealed in aluminum foil, cheese sauce, sautéed onions under a layer of butter, garlic in oil, salted or fermented fish, Jalapeno peppers, potato salad, etc. [6].

Infant botulism results from the colonization of the intestines by spores of *C. botulinum* with subsequent toxin production *in vivo*. After the identification of infant botulism in 1976, a total of 1442 cases were reported to the CDC over next 20 years. Most of these cases were associated with the type B toxins (51%) followed by type A toxin (46%). Approximately 20% of these infants had ingested honey before they developed symptoms [5].

Wound botulism results from production of toxin by *C. botulinum* in a contaminated wound. It is relatively uncommon. Most of the cases were secondary to type A toxin followed by type B toxin. In recent years wound botulism has more often been associated with illicit drug usage, such as sniffing cocaine or injecting so-called "black tar heroin".

Child and adult botulism from intestinal colonization are the least common types. This form of botulism is associated with colonization and botulinum toxin production in the intestines by *C. botulinum* or *C. baratii*.

7.4
Microbiology and Toxicology

Clostridium botulinum is a spore-forming, obligate anaerobic bacillus. In young cultures it is a motile, Gram-positive rod (straight to slightly curved), with oval, subterminal spores (Fig. 7.1). It is approximately 0.5–2.0 μm in width and 1.6–22.0 μm in length. *C. botulinum* is commonly found in soil and aquatic habitats throughout the world. There are four genetically diverse groups of this organism. In addition to *C. botulinum*, two other species of clostridia, viz. *Clostridium baratii* and *Clostridium butyricum*, produce botulinum toxin [7].

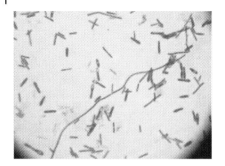 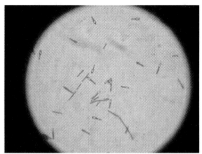

Fig. 7.1
Clostridium botulinum with spores from culture. Courtesy: David
B. Fankhauser, Ph.D., Professor of Biology and Chemistry, University of Cincinnati Clermont College, Batavia, Ohio.

Botulinum toxin, the most poisonous substance known to humans, is a zinc protease enzyme produced by *Clostridium botulinum*. The seven subtypes of botulinum toxin produced by *C. botulinum* are designated with the letters A, B, C, D, E, F, and G. Most described cases of human botulism have resulted from toxin types A, B, and E produced by strains of *C. botulinum*. Human cases of botulism involving toxin types F and E produced by *C. baratii* and *C. butyricum*, respectively, have also been described. Botulism caused by type C and type D botulinum toxins have been described in non-human species only [6].

All clostridial neurotoxins are synthesized as a single inactive polypeptide chain of 150 kDa molecular weight without a leader sequence. They are presumably released from the cells by bacterial lysis [8]. In its natural form the botulinum toxin is a weak toxin. To become fully active, the single-chain molecule must be cleaved by proteolysis to generate a heavy chain (100 kDa) that is linked to a light chain (50 kDa) by a disulfide bond. This dichain form of the molecule causes toxicity and has therapeutic benefit [9].

Botulinum toxin cleaves fusion proteins, which are needed at neuronal vesicles to release acetylcholine into the neuromuscular junction. It binds to receptors on nerve endings with high affinity and penetrates the cell membrane by receptor-mediated endocytosis. The toxin then crosses the endosome membrane by pH-dependent translocation. When it reaches the cytosol, botulinum toxin acts as a zinc-dependent endoprotease to cleave polypeptides that are essential for the exocytosis of neurotransmitters. In the absence of these peptides, nerve impulses can no longer trigger the release of acetylcholine (Fig. 7.2). This blockade of acetylcholine results in failure of neurotransmission along the neuromuscular junction resulting in failure of muscle contractions and flaccid paralysis [9–11]. Studies suggest that release of molecules contained inside exocytic granules and synaptic vesicles at the neuronal endings is mediated by the assembly of a SNARE (soluble *N*-ethylmaleimide-sensitive factor attachment protein receptor) complex formed by the coil-coiling of three proteins: SNAP-25, syntaxin, and VAMP/

synaptobrevin. It seems that the SNARE complexes assemble in rosette-shaped super-complexes for successful acetylcholine release, but there is disagreement on the number of copies of SNARE complexes necessary to mediate exocytosis [12].

LD50, defined as lethal dose for 50% of the exposed population, for botulinum toxin is estimated to be 1 ng kg^{-1} body weight, given parenterally. When given by inhalational route, LD50 for humans is 3 ng kg^{-1} body weight [13]. It has been estimated that one gram of crystalline toxin is sufficient to kill one million people [14].

The botulinum toxin can be inactivated in an environment with unfavorable pH or temperature. The botulinum toxin is stable under acidic conditions (pH 3.5 to 6.5), but the complex dissociates under slightly alkaline conditions and its biological activity is readily inactivated by alkali. Although the spores of *C. botulinum* are relatively heat-resistant, the toxin itself is heat sensitive. Heating at 80 °C for 30 min or 100 °C for 10 min destroys the activity of the toxin [15].

During 1970–1989 Alan Scott used botulinum toxin as an experimental agent in primates followed by its use in humans in an attempt to find a nonsurgical treatment for some forms of strabismus. Botulinum toxin preparations are now available for a variety of therapeutic uses in most countries around the world. The United States Food and Drug Administration (US FDA) has approved botulinum toxin preparations (type A = Botox, type B = Myobloc) for the treatment of cervical dystonia, blepharospasm, spasmodic torticollis, strabismus, glabellar frown lines, and primary axillary hyperhidrosis [13, 16]. They are being studied for treatment of other conditions. In addition, different preparations of botulinum toxin are used for many other off-label therapeutic and cosmetic purposes [17, 18].

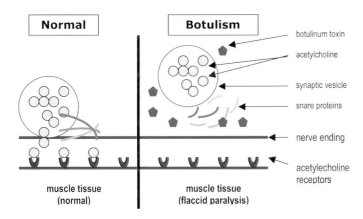

Fig. 7.2
Mechanism of neuromuscular blockade by botulinum toxin. On the left, normal muscle tissue – normal SNARE (soluble *N*-ethylmaleimide-sensitive factor attachment protein receptor) proteins with normal release of acetylcholine at the neuromuscular junction.

On the right, flaccid paralysis by botulinum toxin – SNARE proteins are cleaved by toxin, resulting in inhibition of synaptic vesicle fusion and blockade of acetylcholine release. Adapted, with permission, from Ref. [41].

7.5
Transmission

Botulism can be transmitted in a variety of modes. Two natural forms of botulism, food-borne and intestinal, are food-borne infections. The bacteria multiply in the food or in the intestine, resulting in toxin production. In food-borne botulism, patients ingest contaminated food with preformed toxin. Because the bacteria are ubiquitous, food can become contaminated during food preparation or manufacture. When the contaminated food is stored and processed in a way that allows anaerobic organisms to grow and multiply, the bacteria produce and release toxin as a part of their growth cycle. If food is subsequently heated before consumption, toxin can be destroyed. Food consumed without heating can result in significant toxicity and illness. After consumption, the toxin is absorbed in the gastrointestinal tract and carried to the neuromuscular junctions through the bloodstream.

In contrast, colonization of the intestinal tract after ingestion of spores of *C. botulinum*, and subsequent uninhibited growth of the bacteria can result in toxin production *in vivo*. The intestinal tract of infants often lacks both the protective bacterial flora and the clostridium-inhibiting bile acids found in the normal adult intestinal tract, thus providing suitable conditions for growth of *C. botulinum*. The toxin released in the gut is subsequently absorbed, resulting in infant botulism. A similar mechanism can rarely cause adult botulism after intestinal colonization by botulinum toxin-producing *Clostridia* species. Wound botulism results from direct inoculation of a wound with *C. botulinum* followed by multiplication of bacteria and toxin production in necrotic tissues of the wound [9, 15].

Inhalational botulism can result from a manmade biological weapon with aerosolized botulinum toxin [14]. Studies in monkeys indicate that aerosolized botulinum toxin can be absorbed through the lungs [6].

Three cases of botulism in laboratory workers have been ascribed to inhalation of the toxin [19]. Cases of inadvertent botulism have also been reported after intramuscular administration of the toxin for therapeutic purposes [15].

7.6
Clinical Features

Botulism causes a descending flaccid paralysis. The classic triad of clinical features consists of:
  1. acute, symmetric, descending flaccid paralysis with prominent bulbar palsies;
  2. absence of fever; and
  3. clear sensorium.

Features of bulbar palsy, which include diplopia, dysarthria, dysphonia, and dysphagia, are present in ≥90% of patients. Ocular symptoms and signs consist of blurred vision, ptosis, diplopia, and enlarged pupils with sluggish reaction. The flaccid paralysis progresses in a descending manner with loss of head control, generalized weakness and hypotonia, loss of deep tendon reflexes, airway obstruction, and respiratory muscle paralysis requiring ventilatory support. Death results from airway obstruction and respiratory muscle paralysis. Muscle weaknesses may persist for weeks to months.

In food-borne botulism, the initial symptoms may occur after an incubation period of 12–72 h (range 2 h to 8 days) [14]. Gastrointestinal symptoms, for example abdominal cramps, nausea, vomiting, and diarrhea, are often present. Constipation predominates after the onset of neurological symptoms. The clinical features of adult infectious botulism secondary to intestinal colonization resemble those of food-borne botulism except for the initial gastrointestinal symptoms.

Infant botulism occurs in children under one year of age. It occurs mostly during the first 6 months of life. The clinical syndrome may vary in children. In most cases, constipation may be the first symptom, followed by a progressive descending weakness and poor feeding. The weakness evolves over hours to days. The infant also shows signs of lethargy, difficulty in sucking and swallowing, weak cry, pooled oral secretions, and loss of head control. Neurological examination may reveal hypotonia, ptosis, dilated and sluggish pupils, ophthalmoplegia, weak gag reflex, and dry oral mucosa.

Most cases of wound botulism occur as a consequence of a grossly contaminated wound. These wounds are usually deep and contain necrotic tissue making a suitable anaerobic environment for bacterial growth. The wounds may not always be apparent. The median incubation period is 7 days (range 4–21 days). Many outbreaks in recent years have been related to illicit drug use. Neurological findings are similar to those seen in food-borne botulism [5].

7.7
Diagnosis

Presumptive diagnosis of botulism should be made on the basis of clinical features. A clinical syndrome of descending flaccid paralysis with bulbar palsies in a patient with normal sensorium and no fever should trigger suspicion of botulism.

In clinically suspected cases of botulism, electrophysiological studies can be useful for establishing a presumptive diagnosis. A small evoked muscle action potential in response to a single supramaximal nerve stimulus has been described as the most consistent finding in a clinically affected muscle. Post-tetanic facilitation has also been described in some affected muscles [20]. Single-fiber electromyography (SFEMG) was found to be more sensitive than conventional methods in an outbreak [21].

Botulinum toxin and *Clostridium botulinum* are classified as select agents. Both are regulated in the USA under 42 CFR part 73, titled-"Possession, Use, and Transfer of Select Agents and Toxins". Toxin detection should be performed only by trained individuals at laboratory response network (LRN) reference or higher laboratories. Local or sentinel laboratories are instructed not to perform culture, identify the organism, or attempt to perform toxin analysis [22].

Botulinum toxin assays are available in the state public health laboratories and the CDC. The presence of botulinum toxin in samples of serum, gastrointestinal contents, and suspected source, for example food items, is detected and identified by mouse bioassay. This assay can detect as little as 0.03 ng of the toxin. Mice are usually given intraperitoneal injections of the clinical samples, for example stool or food extracts or other suspect samples. The control mice are injected with a mixture of the sample and the neutralizing antibody to different toxin types. These mice are observed for up to 4 days for development of any signs of botulism, which usually occur within 6 to 24 h.

Fecal and gastric samples should be sent for *C. botulinum* culture. It grows in anaerobic culture media over 7–10 days (range 5–20 days). Acceptable clinical specimens for culture and toxin assays include gastric contents, exudates, tissues, and serum. Postmortem specimens, culture isolates, food samples (solid or liquid), and environmental samples, for example soil and water, can also be sent for identification of the bacteria and its toxin [21, 22].

In case of intentional release of botulinum toxin by aerosolization, the Center for Disease Control and Prevention (CDC) and the American Society of Microbiology (ASM) recommend collection of the specimens [22]:

1. serum
2. feces or return from sterile water enema,
3. food (solid or liquid),
4. environmental or nasal swabs, and
5. gastric aspirate.

Enzyme-linked immunosorbent assay on samples from nasal mucosa taken within 24 h of exposure has been used by the military for detection of aerosolized toxin [24]. Alternative *in vitro* assays including amplified ELISA and PCR are also being developed [5, 25, 26].

A lumbar puncture with normal cerebrospinal fluid results could be useful for excluding other central nervous system infections.

Table 7.2 illustrates the case definitions and diagnostic criteria for surveillance of various forms of botulism as recommended by the CDC.

Tab. 7.2
CDC case definitions and diagnostic criteria for botulism.

Type	Clinical description	Laboratory criteria	Case classification
Food-borne botulism	Results from ingestion of botulinum toxin; diplopia, blurred vision, bulbar palsy, progressive symmetric paralysis	Detection of botulinum toxin in serum, stool, or patients food, or isolation of *C. botulinum* from stool	Probable case: A clinically compatible case with an epidemiological link Confirmed case: A clinically compatible case that is laboratory confirmed, or that occurs among persons who ate the same food as persons who have laboratory-confirmed botulism
Infant botulism	Toxin produced by *C. botulinum* in intestines of infants; constipation, poor feeding, and "failure to thrive", may progress to weakness, impaired respiration, and death	Detection of botulinum toxin in stool or serum, or isolation of *C. botulinum* from stool	A clinically compatible case that is laboratory-confirmed, occurring in a child aged less than 1 year
Wound botulism	Toxin produced by *C. botulinum* that has infected a wound; diplopia, blurred vision, bulbar palsy, progressive symmetric paralysis	Detection of botulinum toxin in serum, or isolation of *C. botulinum* from wound	A clinically compatible case that is laboratory confirmed in a patient who has no suspected exposure to contaminated food and who has a history of a fresh, contaminated wound during the 2 weeks before onset of symptoms
Other botulism (e.g. adult intestinal, inhalational, etc.)	Toxin produced by *C. botulinum* in intestines; diplopia, blurred vision, bulbar palsy, progressive symmetric paralysis	Detection of botulinum toxin in clinical specimen, or isolation of *C. botulinum* from clinical specimen	A clinically compatible case that is laboratory confirmed in a patient aged greater than or equal to 1 year who has no history of ingestion of suspect food and has no wounds

Modified from CDC (1997) Case Definitions for Infectious Conditions under Public Health Surveillance. MMWR 46(RR10):1–55 [40]

7.8
Differential Diagnosis

Other diseases causing muscle weakness should be considered in the differential diagnosis (Table 7.3). The closest of these are Guillain–Barré syndrome (Miller–Fisher variant) and myasthenia gravis, including Lambert–Eaton myasthenic syndrome (LEMS). Tick paralysis in the appropriate setting should also be considered in the differential diagnosis of botulism.

Tab. 7.3
Differential diagnosis of botulism.

Guillain–Barré syndrome (Miller–Fisher variant)
Myasthenia gravis
Lambert–Eaton myasthenic syndrome (LEMS)
Tick paralysis
Poliomyelitis
Encephalitis
Poisonings – atropine, organophosphate, shellfish, mushrooms, etc.

Guillain–Barré Syndrome (GBS) usually causes ascending paralysis, although Miller–Fisher variant may cause descending weakness. Patients with GBS usually have more pronounced cranial nerve involvement, glove and stocking distribution of paresthesias, elevated CSF protein (usually after 1–6 weeks), and abnormal nerve conduction velocity on electromyography (EMG).

Patients with myasthenia gravis show weakness with repetitive actions. They show dramatic response to the edrophonium chloride. EMG shows decrease in muscle action potentials with repetitive stimulation of the neurons.

Tick paralysis can also initially resemble botulism but it presents with ascending paralysis, paresthesias, and usually with cranial nerve involvement. Careful examination may reveal the presence of a tick attached to skin. Patients recover within 24 h of tick removal. EMG shows abnormal nerve conduction velocity, and unresponsiveness to repetitive stimulation.

As shown in Table 7.3, other differential diagnoses include: poliomyelitis, encephalitis, and poisoning with atropine, organophosphates, shellfish, mushrooms, etc.

7.9
Potential as a Biological Weapon

Historical evidence suggests that botulinum toxin has been developed and used as a biological weapon by various countries and terrorist groups in the past. The Japanese fed their prisoners *C. botulinum* cultures in 1930s with fatal effects. During World War II, there were concerns that Germans had developed biological weapons using agents such as anthrax and botulinum toxin, because of this allied troops received botulinum toxoid vaccine [24]. The United States produced botulinum toxin as a potential biological weapon during World War II. Although the US offensive biological weapons program ended before the Biological and Toxin Weapons Convention (BTWC) in 1972, the former USSR conducted research on botulinum toxin as a biological weapon until the early 1990s [27]. A Japanese cult Aum Shinrikyo dispersed aerosols in Tokyo and US military installations in Japan on three occasions between 1990 and 1995 [14]. A United Nations inspection team visiting Iraq after Gulf War (1991) reported that Iraq admitted to having 19,000 L of concentrated botulinum toxin, enough to kill the entire human population three times by inhalation. It was believed that Iraq deployed 600-km range missiles and 100-lb bombs filled with botulinum toxin, aflatoxin, and anthrax spores during 1990 Gulf war [14]. Although extensive searches were subsequently conducted by the inspection teams from the United Nations and United States military, they failed to find any evidence of production of biological warfare agents.

Botulinum toxin can be used in various ways as a weapon. During war, it can be used in military action to immobilize the opponent. It may also be used by terrorists as a lethal weapon to paralyze, to cause death and disability, and to create panic among a large population of civilians. It has been estimated that if used as a point source aerosol in a densely populated area, botulinum toxin can incapacitate or kill 10% of the population within 0.5 km downwind. Inhalational botulism has an incubation period of approximately 72 h [19]. Botulinum toxin can also be used to contaminate food deliberately [14].

7.10
Features of a Botulism Attack

In a suspected botulism outbreak, a careful history about type and source of food, types of activity, and recent travel could be helpful in identification of the source of the outbreak. Features of such a deliberate attack are summarized in Table 7.4. If a large number of individuals develop acute flaccid paralysis with prominent bulbar palsies, it should raise suspicion of deliberate release of botulinum toxin. Affected individuals should be asked about others with similar symptoms. If an outbreak occurs in people living within a common geographic area but they do not have a common dietary exposure, possibility of an aerosol attack should be strongly considered. Similarly, multiple simultaneous outbreaks without a common source or outbreaks of an unusual botulinum toxin type are also suggestive of deliberate attacks. Such unusual toxin types for humans include C, D, F, G, or type E toxins not acquired from an aquatic food [14].

Tab. 7.4
Features of botulism suggesting a deliberate attack.

An outbreak involving a large number of cases of acute flaccid paralysis with prominent bulbar palsies

An outbreak caused by an unusual toxin type (C, D, F, or G) or an outbreak involving type E toxin without an apparent aquatic source

Cases in a common geographic location during the week before symptom onset but without a common food exposure, suggesting aerosol release

Multiple simultaneous outbreaks without a common source

Modified from Ref. [14]

7.11
Management

The two main components in the management of botulism are passive immuni-zation and supportive care. To minimize neurological progression, botulinum antitoxin should be administered as soon as the presumptive clinical diagnosis is made. The US FDA has temporarily suspended the use of trivalent ABE antitoxin until further safety data are available. Currently, two separate equine antitoxins are available through CDC – botulinum antitoxin bivalent for type A and B toxins (licensed by the FDA) and botulism antitoxin type E (an investigational product) [19]. Although one vial of antitoxin is usually sufficient to neutralize food-borne botulism, in cases of intentional exposure, adequacy should be confirmed by retesting patients serum for toxin after administration of a standard initial dose of antitoxin. Children, pregnant women, and immunocompromised patients have been treated safely with equine antitoxin.

Common adverse effects of the antitoxin include urticaria, serum sickness, mild hypersensitivity, and anaphylaxis. A skin test should be performed on all patients before administration of the antitoxin. Patients showing signs of hypersensitivity, for example wheal and flare reaction, should be desensitized with incremental doses of antitoxin administered over 3–4 h [14].

A human antitoxin (human botulism immune globulin, HBIG) is available as an investigational agent for infant botulism [28]. It can be obtained by contacting the California Department of Public Health. A heptavalent (A–G) investigational antitoxin is held by the US Army.

The other important aspect of the management of botulism is adequate suppor-tive care. For airway protection and improved respiratory function patients should be kept in a reverse Trendelenburg position, i.e. head of the bed tilted approx-imately 20–25 degrees above the horizontal level. Assisted ventilation is required for 20% of adult cases and 60% of infants. Attention should be given to adequate fluid and nutritional support.

The role of antibiotics in the treatment of botulism remains uncertain except for treatment of secondary infections complicating botulism. Penicillin G has been recommended for treatment of wound botulism to reduce the bacterial load and toxin production in the wounds. Aminoglycosides and clindamycin are contra-indicated because of their ability to exacerbate neuromuscular blockade [14, 29, 30].

7.12
Prognosis

Mortality rate for food-borne botulism has declined from 60% before 1950 to 15% in the latter half of the 20th century, largely because of better supportive care including improvements in mechanical ventilation and intensive care meas-ures [10]. Early deaths result from a failure to recognize the severity of disease or from secondary pulmonary or systemic infections. Deaths after 2 weeks usually occur from complications related to long-term mechanical ventilation. Fewer than 2% of reported cases of infant botulism result in death. Mortality from wound botulism ranges from 15 to 44% [8, 11, 12].

7.13
Prevention

Most cases of botulism are acquired by ingestion of contaminated food. Appropriate methods for preparation and preservation of food are the most important aspects of prevention of botulism. Immunization and post-exposure prophylaxis are also available for the individuals at high risk. Infection control and decontamination procedures are discussed below.

Most outbreaks of botulism result from improperly preserved canned food prepared at home. Education in preparing and preserving food are important public health measures for controlling botulism. When food is prepared for canning or for preserving for longer duration, it should be cooked at the appropriate temperature and pressure for a time adequate to kill *C. botulinum* spores. For example, use of a pressure cooker can maintain the cooking temperature above the boiling point of water, which is required to destroy the spores. Foods that cannot be heated to such high temperature should be acidified to inhibit the growth of the organism. Incompletely processed foods require appropriate refrigeration. Heating canned foods to 85 °C for 5 min before consumption can destroy heat-labile toxins. Canned food containing botulinum toxin can cause the container to bulge and change the odor of the food. Food-borne botulism in commercial foods has largely been controlled by safe canning and food-manufacturing processes in the United States. Occasional outbreaks can still occur by consumption of commercial foods [5].

7.13.1
Immunization

A pentavalent botulinum toxoid vaccine containing toxoid types A, B, C, D, and E is an investigational vaccine available only for high risk laboratory personnel and the military. It is administered subcutaneously as a 0.5-mL dose at 0, 2, and 12 weeks, with a booster dose at 1 year. To check the adequacy of protection, antitoxin titers are measured every two years after the booster dose. A serum antitoxin titer of 1:16 or 0.15 to 0.30 IU mL^{-1} is considered evidence of adequate immunity. Individuals with lower titers should receive additional doses of the toxoid [33]. Efficacy of the pentavalent botulinum toxoid (ABCDE) was tested in military personnel during the Gulf War in the 1990 s. An immunogenic response was observed in only 28% after the first booster but this response significantly improved to 97% after a second booster [34].

Other vaccines currently under investigation include a tetravalent vaccine (ABEF) developed in Japan, a vaccine using Venezuelan equine encephalitis (VEE) virus vector, and a heavy-chain-based inhalational vaccine [35–37]. A recombinant botulinum vaccine (rBV) A/B developed by DynPort is undergoing first-phase clinical trial in the USA.

7.13.2
Post-exposure Prophylaxis

Botulinum antitoxin used for post-exposure prophylaxis, or rather as an early treatment after exposure to botulinum toxin, has been shown to reduce neurological manifestations. CDC maintains a botulism surveillance system and provides antitoxin through its network within United States. CDC also provides antitoxin to other countries of the eastern hemisphere through its contract with PAHO (Pan American Health Organization) with the exception of Canada, which has its own supply of botulinum antitoxin [38].

7.13.3
Decontamination

Botulinum toxin is heat labile and can be destroyed by heating food and drink to a core temperature of 85 °C for 5 min. All suspected food should be removed and sent to the appropriate public health department for testing. Exposed clothes and skin should be washed with soap and water. Contaminated objects and surfaces should be cleaned with 0.1% hypochlorite solution or with household bleach diluted 1:10. Covering the mouth and nose with clothing provides some protection from aerosolized toxin [14, 27].

7.14
Infection Control

Person-to-person transmission of botulism does not occur. Generally, following standard or universal precautions is sufficient in the healthcare setting. Simple measures, for example using gloves, gowns, goggles, etc., while handling contaminated materials, and washing the hands with soap and water after inadvertent contact with the toxin, should be followed. Because botulism is classified as a reportable select agent, clinicians caring for patients with suspected botulism should immediately contact their local hospital epidemiologist or infection-control practitioner. Subsequently, the local and state health departments should be notified. In USA, CDC has a twenty-four-hour telephone number (770–488–7100) for state health departments to report suspected cases of botulism, to obtain clinical consultation on botulism cases, and to request botulinum antitoxin release. These calls are taken by the CDC Emergency Operations Center [39]. Table 7.2 shows the CDC's case definitions and diagnostic criteria for different forms of botulism, which are used for surveillance of the disease [40].

References

1 Ergbuth FJ, Naumann M. **1999** Historical aspects of Botulinum toxin: Justinus Kerner (1786–1862) and "Sausage Poison". Neurology 53:1850–1853.

2 Erbguth FJ. **2004** Historical notes on botulism, Clostridium botulinum, botulinum toxin, and the idea of the therapeutic use of the toxin. Mov Disord 19 Suppl 8:S2–6.

3 Kreyden OP, Geiges ML, Boni R, et al. **2000** Botulinum toxin: from poison to drug. A historical review. Hautarzt. Oct; 51:733–737.

4 CDC. **2002** Summary of Botulism Cases Reported in 2001, CDC, Atlanta, GA. Available at: http://www.cdc.gov/ncidod/dbmd/diseaseinfo/files/BotCSTE2001.pdf. (Accessed on: 8/12/2005)

5 Centers for Disease Control and Prevention. **1998** Botulism in the United States 1899–1996: Handbook for Epidemiologists, Clinicians, and Laboratory Workers. Atlanta: Centers for Disease Control and Prevention

6 Shapiro RL, Hatheway C, Swerdlow DL. **1998** Botulism in the United States: A clinical and epidemiological review. Ann Intern Med 129:221–228.

7 Suen JC, Hatheway CL, Steigerwalt AG, et al. **1988** Genetic confirmation of identities of neurotoxigenic *Clostridium baratii* and *Clostridium butyricum* implicated as agents of infant botulism. J Clin Microbiol 26:2191–2192.

8 Schiavo G, Rossetto O, Tonello F, Montecucco C. **1995** Intracellular targets and metalloprotease activity of tetanus and botulism neurotoxins. Curr Top Microbiol Immunol 195: 257–274.

9 Simpson LL. **1996** Botulinum toxin: a deadly poison sheds its negative image. Ann Intern Med. 125(7):616–7.

10 Simpson LL, Coffield JA, Bakry N. **1994** Inhibition of vacuolar adenosine triphosphatase antagonizes the effects of clostridial neurotoxins but not phospholipase A2 neurotoxins. J Pharmacol Exp Ther 269:256–62.[Abstract]

11 Schiavo G, Rossetto O, Benfenati F, et al. **1994** Tetanus and botulinum neurotoxins are zinc proteases specific for components of the neuroexocytosis apparatus. Ann N Y Acad Sci. 710:65–75

12 Montecucco C, Schiavo G, Pantano S. (2005) SNARE complexes and neuroexocytosis: how many, how close? Trends Biochem Sci 30(7): 367–72.

13 National Drug Data File (NDDF) from First Data Bank, available on www.medscape.com; accessed on 8/12/2002.

14 Arnon SS, Schechter R, Inglesby T, et al. **2001** Botulinum toxin as a biological weapon: Medical and public health management. JAMA. 285:1059–1070.

15 Nantel AJ. **1999** Clostridium botulinum: International Programme on Chemical Safety Poisons Information Monograph 858. WHO

16 Carruthers A, Carruthers J. **2001** Update on the botulinum neurotoxins. Skin Therapy Lett. 6:1–2.

17 Scott AB. **2004** Development of botulinum toxin therapy. Dermatol Clin Apr; 22(2):131–3, v

18 Carruthers A. **2003** History of the clinical use of botulinum toxin A and B. Clin Dermatol Nov–Dec; 21(6):469–72.

19 Holzer VE. **1962** Botulism from inhalation. Med Klin 57:1735–1738.

20 Cherington M. **1998** Clinical spectrum of botulism. Muscle Nerve. 21:701–710.

21 Padua L, Aprile I, Monaco ML, et al. **1999** Neurophysiological assessment in the diagnosis of botulism: usefulness of single-fiber EMG. Muscle Nerve. 22:1388–1392.

22 Shapiro DS, Weissfield AS. Sentinel Laboratory guidelines for suspected agents of bioterrorism: Botulinum Toxin. American Society for Microbiology. Available at: http://www.bt.cdc.gov/agent/botulism/index.asp. Accessed on 8/14/2005.

23 CDC, ASM, APHL.**2001** Basic Protocols for Level A Laboratories for Botulinum Toxin Dec 1–13.

24 Middlebrook JL, Franz DR.**1997** Botulinum Toxins. In: Textbook of Military Medicine: Medical Aspects of Chemical and Biological Warfare. The Surgeon General 643–654.

25 Franciosa G, Ferreira JL, Hatheway CL.**1994** Detection of type A, B, and E botulism neurotoxin genes in *Clostridium botulinum* and other *Clostridium* species by PCR: evidence of unexpressed type B toxin genes in type A toxigenic organisms. J Clin Microbiol 32:1911–1917.

26 Ferreira JL, Eliasberg SJ, et al. **2001** Detection of preformed type A botulinum toxin in hash brown potatoes by using the mouse bioassay and modified ELISA test. J AOAC Int 84:1460–1464.

27 IDSA. **2003** Bioterrorism Information Resources: Botulinum Toxin. Infectious Disease Society of America, Alexandria, VA, USA. Last updated: May 27, 2003. Available at: http://www.idsociety.org/. Accessed on August 30, 2003.

28 Shen WP, Felsing N, Lang D, et al.**1994** Development of infant botulism in a 3-year-old female with neuroblastoma following autologous bone marrow transplantation: potential use of human botulism immune globulin. Bone Marrow Transplant 13:345–347

29 Santos JI, Swensen P, Glasgow LA. **1981** Potentiation of *Clostridium botulinum* toxin by aminoglycoside antibiotics: clinical and laboratory observations. Pediatrics 68:50–54.

30 Schulze J, Toepfer M, Schroff KC, et al. **1999** Clindamycin and nicotinic neuromuscular transmission. Lancet 354:1792–1793.

31 Werner SB, Passaro K, McGee J, et al. **2000** Wound botulism in California, 1951–1998: recent epidemic in heroin injectors. Clin Infect Dis 31:1018–1024.

32 Merson MH, Dowell VR. **1973** Epidemiological, clinical, and laboratory aspects of wound botulism. N Engl J Med 289:1005–1010.

33 CDC. Drug Service: General information. Available at: http://www.cdc.gov.

34 Pittman PR, Hack D, et al. **2002** Antibody response to a delayed booster dose of anthrax vaccine and botulinum toxoid. Vaccine 20:2107–2115.

35 Torii Y, Tokumaru Y, et al. **2002** Production and immunogenic efficacy of botulinum tetravalent (A,B,E,F) toxoid. Vaccine 20:2556–2561.

36 Lee JS, Pushko P, et al. **2001** Candidate vaccine against botulinum neurotoxin serotype A derived from a Venezuelan equine encephalitis virus vector system. Infect Immun 69:5709–15.

37 Park JB, Simpson LL. **2003** Inhalational poisoning by botulinum toxin and inhalation vaccination with its heavy-chain component. Infect Immun 1147–1154.

38 Shapiro RL, Hatheway C, Becher J, et al. **1997** Botulism surveillance and emergency response: a public health strategy for a global challenge. JAMA 278:433–435.

39 CDC. **2003** Notice to Readers: New Telephone Number to Report Botulism Cases and Request Antitoxin. MMWR 52(32);774.

40 CDC. **1997** Case Definitions for Infectious Conditions under Public Health Surveillance. MMWR May 2;46(RR10):1–55.

41 Koirala J, Basnet S. **2004** Botulism, Botulinum Toxin and Bioterrorism. Infect Med 21(6):284–290.

8
Tularemia: Natural Disease or Act of Terrorism

Janak Koirala

8.1
Introduction

Tularemia is a zoonosis caused by *Francisella tularensis*. It is also commonly known as rabbit fever, deer-fly fever, and Ohara's disease (Japan). Its distribution is limited to the Northern hemisphere only, with remarkable absence from the Southern hemisphere. Animal reservoirs of *F. tularensis* are usually rodents and lagomorphs. It is transmitted by insect vectors, such as ticks and flies. In human, various clinical forms of the disease have been described including skin ulcers, ocular infection, lymphadenitis, and pneumonic, septicemic, and typhoidal forms. It typically causes a febrile illness accompanied by regional or generalized lymphadenopathy. The rapid transmissibility and high infectivity of *F. tularensis* have led to its designation as a potential weapon of bioterrorism. It is regarded as a "category A" agent.

8.2
History

Tularemia was first described in Japan in 1918 as a hare-associated illness. In 1911, George Walter McCoy described it for the first time in the USA as a plague-like disease of rodents (ground squirrel). Later, it was recognized to cause a severe, fatal, human illness. The causative organism of tularemia was isolated in 1912. Edward Francis established the cause of deer-fly fever, proved deer-fly was the vector, and named the disease "tularemia" as the original work took place in Tulare County, CA, USA. The causative organism was originally named *Bacterium tularense* by Edward Francis. Later, its name was changed to *Pasteurella tularensis,* and subsequently to *Francisella tularensis* after Edward Francis to recognize his lifetime contributions to defining and understanding the disease. Some of his important

Bioterrorism Preparedness. Edited by Nancy Khardori
Copyright © 2006 WILEY-VCH Verlag GmbH & Co. KGaA, Weinheim
ISBN: 3-527-31235-8

contributions included improved laboratory methods, development of serology, and identification of other reservoirs. He also emphasized possible risks of acquiring tularemia by laboratory workers and consumers exposed to infected animals and animal products [1–4].

8.3
Microbiology

Francisella tularensis is the more commonly isolated organism of the two recognized species of the genus *Francisella*. The other species, *Francisella philomiragia*, is rarely isolated from clinical specimens [5].

F. *tularensis* is a small, obligately aerobic, pleomorphic, nonmotile, gram-negative coccobacillus. It takes a bipolar staining with Gram or Giemsa stains, which gives it a coccoidal appearance (Fig. 8.1). Its growth requires culture media enriched with cysteine [6, 7]. The virulence of the F. *tularensis* is not well understood, but it has been associated with its capsule and citrulline ureidase activity [8].

Classification of F. *tularensis* has been confusing and different names for subspecies have been described in the literature. On the basis of their virulence and biochemical and epidemiological features, four subspecies of F. *tularensis* have recently been identified (Table 8.1) – F. *tularensis* sub. *tularensis* (also known as type A), F. *tularensis* sub. *holarctica* (also known as *palaearctica* or type B), F. *tularensis* sub. *mediaasiatica*, and F. *tularensis* sub. *novicida* [6]. Their genetic analysis using genomic deletion events and single-nucleotide variations suggested a vertical descent of the four subspecies of *Francisella tularensis*. This analysis also indicated that the highly virulent subspecies, F. *tularensis* sub. *tularensis* (type A), appeared before the less virulent form, F. *tularensis* sub. *holarctica* (type B) [9].

Tab. 8.1
Classification of *Francisella tularensis*.

Biovar	Distribution	Virulence
F. *tularensis*, sub. *tularensis* (type A)	North America	High
F. *tularensis*, sub. *holarctica* (type B)	Asia, Europe, N. America	Low
F. *tularensis*, sub. *Mediaasiatica*	Central Asian Republics of the former USSR	Moderate
F. *tularensis*, sub. *Novicida*	North America	Low

The complete genome sequence of a highly virulent isolate of F. *tularensis* (1,892,819 bp) has been reported [10]. Several virulence-associated genes were found in a putative pathogenicity island with duplicates in the genome. It has been shown that horizontal movement of a genes cluster in the pathogenicity

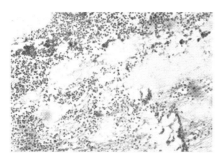

Fig. 8.1
Gram stain of *Francisella tularensis* from a blood culture (Courtesy: Joan Barenfanger MD, Memorial Medical Center, Springfield, Illinois, USA).

island (PI) enables different species of bacteria to gain the ability to invade cells. Similarly, the *Francisella* pathogenicity island (FPI) is required for growth inside macrophages and for virulence [11]. Previously uncharacterized genes encoding type IV pili, a surface polysaccharide and iron-acquisition system were also revealed in the sequence analysis. The genome was found to be rich in insertion sequence (IS) elements, including IS630 Tc-1 mariner family transposons. These IS elements are not usually present in prokaryotes. The biosynthetic pathways of *F. tularensis* are not complete, which forces them to become obligate host-dependent organisms in their natural life cycle [10]. The four subspecies (or biotypes) of *F. tularensis* have greater than 95% DNA sequence identity [11].

8.4
Epidemiology

Tularemia is almost entirely a rural disease with a few exceptions of outbreaks in urban and suburban areas. It is a disease limited to the Northern hemisphere with remarkable absence from the Southern hemisphere. In North America it is prevalent in most of the states of the USA except Hawaii, with the highest rates in Missouri, Arkansas, Oklahoma, and South Dakota. According to the CDC reports the annual incidence of tularemia in USA is estimated to be 0.05 cases per 100,000 population [12]. Its incidence increases during summer months (June through August); this is associated with increased outdoor activity and a greater number of tick-related cases. Tularemia has also been reported from several countries in Europe (Spain, Sweden, Finland, Russia, Czech Republic, Kosovo, and Slovakia), where mostly *F. tularensis* sub. *holoarctica* (type B), also known as *F. tularensis* sub. *palaearctica*, is prevalent. *F. tularensis* sub *holoarctica* and *F. tularensis* sub. *mediaasiatica* have also been reported from Asia (Japan, Turkey, Kazakhstan, Afghanistan, and the central states of the former USSR). No indigenous cases of tularemia have been reported from South America, Africa, and Australia.

The reservoirs for *F. tularensis* are small mammals including voles, mice, water rats, squirrels, rabbits, and hares. They have also been recovered from water, soil, and vegetation. Transmission usually occurs as a result of insect bites, handling infectious animal tissues or fluids, contact or ingestion of contaminated materials, or inhalation of infective aerosols. Arthropods such as ticks, flies, and mosquitoes serve as the common vectors of transmission; animal bites or direct contact with infected animals can also result in the transmission of the infection. A contaminated environment can also serve as a vehicle for transmission of the infection. There have been no documented cases of human-to-human transmission. Hunters, meat handlers, farmers, and laboratory workers are regarded as the high-risk groups for tularemia exposure.

Numerous outbreaks have been reported from various parts of the world. The largest recorded airborne outbreak occurred in Sweden in 1966–67, when *F. tularensis* biovar palaearctica (type B) affected 600 farmers [13].

8.5
Pathogenesis

After successful entry through the skin and mucous membranes, gastrointestinal tract, or respiratory tract, *F. tularensis* multiplies in the local tissue. It subsequently spreads into the regional lymph nodes, multiplies in macrophages, and disseminates to the organs. Suppurative lesions are formed at the site of inoculation; these later change into granulomas with central necrosis. Inhalational exposure to *F. tularensis* may result in bronchiolitis, bronchopneumonia, and hemorrhagic inflammation of the airways, leading to pleuritis and hilar adenopathy.

8.6
Clinical Features

Tularemia has an incubation period of 2–10 days. Patients usually develop abrupt onset of fever with chills and rigors, headache, body aches, upper respiratory symptoms, and pulse–temperature dissociation (i.e. failure of rise in pulse rate as expected for an increased temperature). Different clinical forms of tularemia have been described (Table 8.2).

Tab. 8.2
Clinical forms of tularemia.

Ulceroglandular tularemia
Glandular tularemia
Oculoglandular tularemia
Oropharyngeal tularemia
Pneumonic tularemia
Typhoidal tularemia
Septicemic tulremia

---

1. **Ulceroglandular tularemia:** Most (45–80%) patients with tularemia present with a cutaneous ulcer with regional lymphadenopathy. A local painful cutaneous lesion, for example a papule or ulcer, is usually found at the site of inoculation. The papule changes into an ulcer in a few days. Patients develop tender regional lymphdenopathy simultaneously with, before, or after the skin lesions. Cervical and occipital adenopathies are more common in children whereas inguinal nodes are more often involved in adults, who also have fever, chills, myalgias, and headache. Other skin lesions, for example erythema nodosum or erythema multiforme, may also occur.

2. **Glandular tularemia:** Patients with glandular tularemia have regional lymphadenopathy without any skin lesions. It is the second most common variety in the USA and the most common form seen in Japan. They have similar presentations to the ulceroglandular form except for the skin lesions. They also develop similar constitutional symptoms, for example fever, chills, generalized myalgia, etc.

3. **Oculoglandular tularemia:** The oculoglandular form is seen in fewer than 5% of cases. It occurs as a result of the inoculation of bacteria through conjunctiva. Patients develop excessive tearing, lid swelling, and painful, purulent conjunctivitis. They may have conjunctival ulcers. They also develop periauricular, submandibular and cervical lymphadenopathy, which are usually tender. Complications of oculoglandular tularemia include corneal ulcers, dacryocystitis, and suppuration of lymph nodes, but blindness is quite rare.

4. **Oropharyngeal tularemia:** More common in children, this form is acquired from contaminated food and water. Patients develop stomatitis, pharyngitis, or tonsillitis with cervical lymphadenopathy. Patients usually complain of a severely sore throat. Examination of pharynx shows exudates, ulcers, and, in some patients, a pharyngeal membrane that may be confused with diphtheria.

5. **Pneumonic:** Some patients may develop a pleuropulmonary disease with a pneumonia-like illness which may result from direct inhalation of the bacteria or as a result of disseminated infection. Patients usually present with minimally productive or nonproductive cough, fever, and pleuritic chest pain. Pleural fluid is exudative and predominantly lymphocytic. Lung biopsy may show granulomas.

6. **Typhoidal tularemia:** Patients with a systemic infection may develop a febrile illness with or without signs of a local disease. Patients usually present with fever, chills, headache, myalgia, nausea, and vomiting. The source of infection is usually not obvious. Patients do not usually have prominent lymphadenopathy or hepatosplenomegaly.

7. **Septicemia tularemia:** This is the most severe form of tularemia, with bacteremia and systemic infection. Patients usually present with nonspecific fever, abdominal pain, diarrhea, and vomiting, with the patient appearing confused. This may progress rapidly to septic shock, disseminated intravascular coagulation (DIC), adult respiratory distress syndrome (ARDS) and death.

8.7
Laboratory Diagnosis

When a case of tularemia is suspected the patient's respiratory and blood samples should be collected and the laboratory should be alerted promptly about the possibility of *F. tularensis*, because laboratories processing these specimens must follow BSL-2 microbiology laboratory procedures [6, 7, 14].

F. tularensis takes bipolar staining with Gram or Giemsa stains, giving it the appearance of small, Gram-negative coccobacilli (Fig. 8.1). It can be grown on cysteine-enriched culture medium from patient samples such as pharyngeal secretions, other respiratory secretions, and also from blood [7]. It produces very scant or no growth on sheep blood agar, but forms grayish white colonies on chocolate agar. Characteristic features of *F. tularensis* include oxidase negative, weak catalase positive, beta-lactamase positive, satellite negative, and urease positive. Identification should be performed in a biosafety cabinet. Antimicrobial susceptibility is performed using the E-test. *F. tularensis* should not be identified with commercial systems because this can generate aerosols. Automated commer-

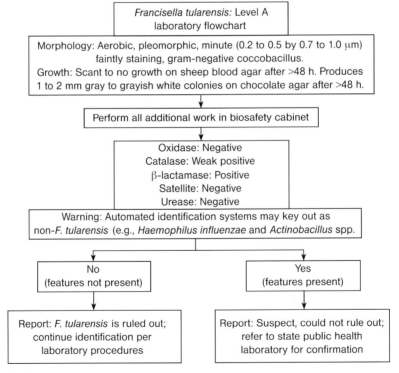

Fig. 8.2
Flow chart for laboratory identification of *F. tularensis*. Adapted from CDC, ASM, APHL Basic Protocols for level A laboratories for the presumptive identification of F. tularensis. Ref. [7].

cial systems may also lead to misidentification of *F. tularensis*. The most common misidentifications are *Haemophilus influenzae* (satellite or XV positive) and *Actinobacillus* species (beta-lactamase negative) [7]. A flow chart developed to help laboratories to identify *F. tularensis* is shown in Fig. 8.2.

A direct fluorescent antibody (DFA) stain, polymerase chain reaction (PCR), or antigen detection tests can be used for rapid identification. DFA is usually recommended as the preferred method. Rapid PCR-based methods have been described for identification of species and subspecies of *Francisella* [5].

For serological diagnosis of tularemia, a microagglutination assay can be used to detect antibodies, beginning 10 days from the onset of infection. A fourfold rise in titer between acute phase and convalescent phase, or a single titer of >1:128 are regarded as diagnostic. Virulence testing and molecular genetic characterizations of *F. tularensis* are also available in specialized laboratories [7, 15]. More recently, a novel enzyme-linked immunosorbent assay (ELISA) and a confirmatory Western blot (WB) to detect human antibodies against *Francisella tularensis* have been evaluated in Germany with sensitivity and specificity for both assays in the range 97–100 % [16].

Histopathology of tissue samples of infected skin, lungs, lymph nodes, spleen, liver, and kidneys show acute suppurative necrosis followed by granulomatous reactions. In a recently reported series, frequent histolopathological findings of irregular microabscesses and granulomas in liver, spleen, kidney, and lymph nodes, and necrotizing pneumonia have been described. Some unusual cases were seen with suppurative leptomeningitis and gastrointestinal ulcers. PCR analysis for a 211-bp fragment of the *F. tularensis* lipoprotein gene on tissue samples was found to be 93 % sensitive [17].

8.8
Radiology

A chest radiograph of patients with the pneumonic form of tularemia may show peribronchial infiltrates, bronchopneumonia of one or more lobes, pleural effusion, and hilar lymphadenopathy. Pulmonary infiltrates on chest X-ray may range from small discrete pulmonary lesions to scattered granulomatous lesions of lung parenchyma or pleura.

8.9
Potential as Biological Weapon

Francisella tularensis is a highly infectious intracellular pathogen which can be aerosolized. Doses as low as 25 colony-forming units can cause a debilitating or fatal disease [18].

F. tularensis was studied as a potential warfare agent in Japan during 1932–1945 and later in the USA in the 1950 s and 1960 s. Russia is believed to have developed strains resistant to antibiotics and vaccines [14, 19, 20]. According to the WHO expert committee's estimates, aerosol dispersal of 50 kg virulent *F. tularensis* over a metropolitan area with five million inhabitants would result in 250,000 incapacitating casualties, including 19,000 deaths [21]. An aerosol release of this kind in a densely populated area would result in a febrile illness in 3–5 days followed by pleuropneumonitis and systemic infection. Tularemia has slower disease progression and a lower case fatality rate than anthrax or plague, but the illness would be expected to persist for several weeks with relapses. The virulence and resistance to antibiotics can be potentially enhanced to transform the bacteria into a more lethal agent [7, 22]. According to a CDC estimates, an aerosol attack with tularemia would cost \$5.4 billion per 100,000 persons exposed [23].

In 2001 the US Working Group on Civilian Biodefense recommended a set of guidelines for considering diagnosis of tularemia as a potential act of bioterrorism [14].

As shown in Table 8.3, sudden onset of acute febrile illness leading to pharyngitis, bronchiolitis, pneumonia, pleuritis, and hilar lymphadenopathy should raise suspicion of tularemia. If an outbreak of tularemia occurs at an epidemic level in an unusual setting, it should raise the suspicion of bioterrorism. Some examples include point-source outbreak in an urban/nonagricultural setting, unexpected severe respiratory illness in otherwise healthy persons, and risk related to degree of exposure and not to age or other risk factors.

Tab. 8.3
Factors suggesting diagnosis of tularemia as a potential act of bioterrorism.

Clinical suspicion of inhalational tularemia:	Sudden onset of acute febrile illness progressing to pharyngitis, bronchiolitis, pneumonitis, pleuritis, and hilar lymphadenitis, which may lead to sepsis and septic shock
Epidemic setting:	Point-source outbreak in an urban/nonagricultural setting; unexpected severe respiratory illness in otherwise healthy persons, risk related to degree of exposure with no difference in susceptibility by age or other risk factors

Adapted from Ref. [14]

8.10
Diagnostic Criteria

As a tool for the surveillance of tularemia, the US Center for Disease Control and Prevention (CDC) has recommended criteria for presumptive and confirmatory diagnosis of tularemia [24]. A presumptive diagnosis of tularemia should be made if one of the following two diagnostic tests gives a positive result:

1. detection of elevated serum antibody titers to *F. tularensis* in a patient with no history of immunization against tularemia, or
2. confirmed detection of *F. tularensis* on DFA assay.

A confirmatory diagnosis of tularemia is established on the basis of either a positive culture result or serological diagnosis using one of the following two criteria:

1. isolation and identification of *F. tularensis*, or
2. a fourfold rise in antibody titer between acute and convalescent phase sera by Microagglutination Assay (Table 8.4).

Tab. 8.4
Diagnostic criteria for tularemia.

Presumptive diagnosis if one of the two criteria met:	Detection of elevated serum antibody titers to *F. tularensis* in a patient with no history of immunization against tularemia, or
	Confirmed detection of *F. tularensis* by DFA assay
Confirmatory diagnosis if one of the two criteria met:	Isolation and identification of *F. tularensis*, or
	A fourfold rise in antibody titer between acute and convalescent phase serum by microagglutination assay

Adapted from Ref. [24]

8.11
Treatment

Aminoglycosides are the preferred antimicrobial agents for treatment of tularemia. Either streptomycin (1 g twice daily, intramuscular) or gentamicin (5 mg kg^{-1} day^{-1}, intramuscular or intravenous) can be used for 10 days. Ciprofloxacin (500 mg twice daily, orally) can be used as an alternative agent, although only in-vitro and animal data are available to support its efficacy. Ciprofloxacin is not FDA approved for tularemia. Other alternatives include tetracyclines or chloramphenicol, which are bacteriostatic agents and may result in a higher rate of relapse. Patients receiving these drugs should be treated for 14–21 days [14, 15]. Pregnant women and children should be treated in the same manner as adults, i.e. using gentamicin or ciprofloxacin (Table 8.5).

Tab. 8.5
Recommended antibiotics for treatment of individuals with
tularemia.

Preferred agents:	Aminoglycosides for 10 days	
	Streptomycin	1 g twice daily, IM
	Gentamicin	5 mg kg^{-1} day^{-1}, IM or IV
Alternative agents:	Ciprofloxacin	500 mg orally twice daily for 10 days (only in-vitro and animal data, not approved for human tularemia)
	Doxycycline	100 mg orally twice daily for 14–21 days
	Tetracycline	500 mg every six hours for 14–21 days
	Chloramphenicol	500 mg every six hours for 14–21 days (tetracyclines or chloramphenicol may result in higher relapse rates)
Children:	Same as adults with dose adjustments based on weight	
Pregnancy:	Gentamicin or ciprofloxacin (benefits outweigh risks to both children and pregnancy)	

In an outbreak or when there are mass casualties, i.e. when the number of casualties may exceed the ability of healthcare system to administer intravenous medication, oral agents should be used (Table 8.6). For adults, doxycycline, 100 mg, taken twice daily, or alternatively ciprofloxacin, 500 mg, taken twice daily are recommended for 14 days. Similarly, children should be given doxycycline, 2.2 mg kg^{-1} twice daily, or ciprofloxacin, 15 mg kg^{-1} twice daily, for the same duration [14].

Tab. 8.6
Recommended antibiotics for a large outbreak of tularemia.

Treat with oral agents for 14 days		
Doxycycline:	Adults	100 mg, PO, twice daily
	Children (<45 kg)	2.2 mg kg^{-1}, PO, twice daily
Ciprofloxacin:	Adults	500 mg, PO, twice daily
	Children	15 mg kg^{-1}, PO, twice daily

8.12
Prevention

8.12.1
Immunization

A live attenuated vaccine is available for laboratory workers as an investigational new drug currently under FDA review. During the 1960 s the same live attenuated vaccine reduced inhalational tularemia in laboratory workers from 5.7 cases to 0.27 cases per 1000 person-years [25]. Complete sequencing of the *F. tularensis* (Schu 4) genome and antigen mapping of live vaccine strain (LVS) were reported in 2000 [26]. After the increased threat of bioterrorism in the late 1990 s, the United State Department of Defense Joint Vaccine Acquisition Program (JVAP) awarded a contract to the DynPort Vaccine Company for development of vaccines against potential biothreat agents, including *F. tularensis*. DynPort produced a new lot of tularemia live vaccine strain (LVS) suitable for human clinical trials and obtained the US Food and Drug Administrations approval in 2004 as a NIAID investigational new drug. A phase 1 clinical trial was begun in early 2005 at Baylor University, Texas, USA [27].

8.12.2
Post-exposure Prophylaxis

Persons exposed to *F. tularensis*, and within the incubation period, should receive doxycycline or ciprofloxacin for 14 days [14]. Alternatively, streptomycin or gentamicin can be administered for the same duration.

8.13
Infection Control

Isolation is not required for patients with tularemia, because human-to-human transmission of *F. tularensis* does not occur. Standard and universal precautions for blood and body fluids should suffice.

Microbiology laboratory personnel should be alerted whenever there is suspicion of a case of tularemia, because they are required to follow BSL-2 precautions for handling these organisms. Procedures potentially resulting in aerosol or droplet production need BSL-3 laboratory conditions. Autopsy procedures which are likely to cause aerosols, e.g. bone sawing, should be avoided. An exposed person should wash the exposed area or whole body with soap and plenty of water. In the event of environmental contamination, e.g. a laboratory spill, etc., 10 % bleach solution can be used for spraying or cleaning. Further decontamination or cleaning can be performed with 70% alcohol for 10 min after using bleach. Standard chlorine level in municipal water should be sufficient to protect against the waterborne disease [14].

8.14
Reporting to the Public Health System

Tularemia is a reportable disease and regarded as one of the "category A" agents of bioterrorism. According to the CDC, "category A" agents include high-priority organisms that pose a risk to national security because they can be easily disseminated or transmitted from person to person, result in high mortality rates, and have the potential for major public health impact, might cause public panic and social disruption, and require special action for public health preparedness.

The Working Group on Civilian Biodefense recommended a set of steps for reporting cases of tularemia [14]. Clinicians caring for patients with suspected tularemia should immediately contact hospital epidemiologist or infection-control practitioner and local or state health departments. If the local and state health departments are unavailable, contact the Centers for Disease Control and Prevention. Updated information on reporting can be found on the CDC website: http://www.cdc.gov/.

The reporting guidelines for microbiology laboratories are published in the Laboratory Protocol developed by CDC, ASM, and APHL [7]. If *F. tularensis* is suspected by the physician, Level A laboratories should consult with state public health laboratory director (or designate) before or concurrent with testing. They should immediately notify the state public health laboratory director (or designate) and the state public health department epidemiologist or health officer if *F. tularensis* cannot be ruled out and a bioterrorist event is suspected. The state public health laboratory and state public health department notifies local FBI agents as appropriate. The physician and infection-control personnel should be informed immediately, according to internal policies of the laboratory, if *F. tularensis* cannot be ruled out. The original specimens should be preserved pursuant to a potential criminal investigation and possible transfer to an appropriate Laboratory Response Network (LRN) laboratory. FBI and state public health laboratory/state public health department will coordinate transfer of the isolates/specimens to a higher-level LRN laboratory as appropriate. Chain of custody documentation should be started as appropriate. Guidance from the state public health laboratory should be obtained as appropriate, for example requests from local law enforcement or other local government officials. If *F. tularensis* is ruled out, efforts should be made to identify the organism using established microbiological procedures.

References

1 Francis E **1925** Tularemia. JAMA 84:1243–1250.
2 McCoy GW **1911** A plague-like disease of rodents. Public Health Bull. 43:53–71.
3 McCoy GW, Chapin CW **1912** *Bacterium tularense*, the cause of a plague-like disease of rodents. Public Health Bull. 53:17–23.
4 Evans ME, Friedlander AM **1997** Tularemia. In: Sidell FR, Takafuji ET, Franz DR, eds. Textbook of Military Medicine: Medical Aspects of Chemical and Biological Warfare. Office of the Surgeon General, Washington, DC 503–512.
5 Johansson A, Ibrahim A, Goransson I, et al. **2000** Evaluation of PCR-based methods for discrimination of *Francisella* species and subspecies and development of a specific PCR that distinguishes the two major subspecies of *Francisella tularensis*. J Clin Microbiol 38(11):4180–4185.
6 *Francisella. 2000* In: Forbes BA, Sahm DF, Weissfeld AS, eds. Bailey and Scotts Diagnostic Microbiology. Mosby, St. Louis, MO 2002:495–497.
7 CDC, ASM, APHL Basic Protocols For Level A Laboratories for the Presumptive Identification of *Francisella Tularensis*. Dec. 2001; 1–13.
8 Tärnvik A **1989** Nature of protective immunity to *Francisella tularensis*. Rev Infect Dis 11:440–450.
9 Svensson K, Larsson P, Johansson D, et al. **2005** Evolution of subspecies of *Francisella tularensis*. J Bacteriol 187(11):3903–3908.
10 Larsson P, Oyston PC, Chain P, et al. **2005** The complete genome sequence of Francisella tularensis, the causative agent of tularemia. Nat Genet 37(2):153–159.
11 Nano FE, Zhang N, Cowley SC, et al. **2004** A Francisella tularensis pathogenicity island required for intramacrophage growth. J Bacteriol 186(19):6430–6436.
12 CDC **2002** Tularemia – United States, 1990–2000. MMWR 51(09):182–184.
13 Dahlstrand S, Ringertz O, Zetterberg B **1971** Airborne tularemia in Sweden. Scand J Infect Dis 3:7–16.
14 Dennis DT, Inglesby TD, Henderson DA, et al. **2001** Tularemia as a biological weapon: medical and public health management. JAMA 285:2763–2773.
15 Miller JM **2001** Agents of bioterrorism. Preparing for bioterrorism at the community health care level. Infect Dis Clin North Am 15(4):1127–1156.
16 Schmitt P, Splettstosser W, Porsch-Ozcurumez M, et al. **2005** A novel screening ELISA and a confirmatory Western blot useful for diagnosis and epidemiological studies of tularemia. Epidemiol Infect 133(4):759–766.
17 Lamps LW, Havens JM, Sjostedt A, et al. **2004** Histologic and molecular diagnosis of tularemia: a potential bioterrorism agent endemic to North America. Mod Pathol. 17(5):489–495.
18 Oyston PC, Sjostedt A, Titball RW **2004** Tularaemia: bioterrorism defence renews interest in Francisella tularensis. Nat Rev Microbiol 2(12):967–978.
19 Harris S **1992** Japanese biological warfare research on humans: a case study of microbiology and ethics. Ann N Y Acad Sci 666:21–52.
20 Christopher GW, Cieslak TJ, Pavlin JA, et al. **1997** Biological warfare: a historical perspective. JAMA 278:412–417.
21 WHO **1970** Health Aspects of Chemical and Biological Weapons. World Health Organization, Geneva, Switzerland 1970:105–107.

22 Pavlov VM, Mokrievich, Volkovoy K **1996** Cryptic plasmid pFNL10 from *Francisella novicida*-like F6168: the base of plasmid vectors for *Francisella tularensis*. FEMS Immunol Med Microbiol 13:253–256.

23 Kaufmann AF, Meltzer MI, Schmid GP **1997** The economic impact of a bioterrorist attack: are prevention and post-attack intervention programs justifiable? Emerg Infect Dis 3:83–94.

24 CDC **1997** Case definitions for infectious conditions under public health surveillance. Centers for Disease Control and Prevention. MMWR Recomm Rep. 46(RR-10):1–55.

25 Burke DS **1977** Immunization against tularemia: analysis of the effectiveness of live *Francisella tularensis* vaccine in prevention of laboratory-acquired tularemia. J Infect Dis 135(1):55–60.

26 Karlsson J, Prior RG, Williams K, et al. **2000** Sequencing of the Francisella tularensis strain Schu 4 genome reveals the shikimate and purine metabolic pathways, targets for the construction of a rationally attenuated auxotrophic vaccine. Microb Comp Genomics 5(1):25–39.

27 DVCs Biodefense Products: Biodefense product news and updates. http://www.csc.com/mms/dvc/en/mcs/mcs365/displayMcs.jsp?id=666. Accessed on 8/3/2005.

9

Viral Hemorrhagic Fevers:
Differentiation of Natural Disease from Act of Bioterrorism

James M. Goodrich

9.1

Introduction

There has been increasing interest in biological agents as possible terrorist weapons. Hemorrhagic fever viruses have been mentioned as potential biological weapons. Unlike smallpox, these viruses commonly cause infection in different parts of the world so presentation of a case with the appropriate travel or work history would not automatically trigger a question about a bioterrorism event. All of these viral infections must, however, be reported immediately to the appropriate agencies. For some of these viruses, we are starting to understand more about the molecular biology and pathogenesis of disease, because of human outbreaks and use of animal models. The greatest danger is from aerosol release.

Viral hemorrhagic fever viruses (HVF) form four major families – the Filoviridae, Arenaviridae, Bunyaviridae, and Flaviviridae [1]. These viruses are usually transmitted to man via infected animals and/or arthropod vectors. Their clinical course may vary, but they are all capable of causing a hemorrhagic fever syndrome. All are enveloped, single-stranded, RNA viruses (Table 9.1). Table 9.2 summarizes these viruses and other viruses of interest covered in this chapter.

Tab. 9.1
Virology of hemorrhagic fever viruses [1]

Family	Morphology	Envelope	Genome size (kbp)	Genome	Segmented
Filoviridae	Filamentous	Yes	19	Single-stranded RNA (−)	Yes
Arenaviridae	Spherical	Yes	11	Single-stranded RNA (±)[a]	Yes
Bunyaviridae	Spherical	Yes	11–19	Single-stranded RNA (−)	Yes
Flaviviridae	Isometric	Yes	10–12	Single-stranded RNA (+)	No

a Ambisense RNA (positive and negative stranded)

Bioterrorism Preparedness. Edited by Nancy Khardori
Copyright © 2006 WILEY-VCH Verlag GmbH & Co. KGaA, Weinheim
ISBN: 3-527-31235-8

Tab. 9.2
Viral hemorrhagic fever and other viruses of interest

Family/virus	Disease	Normal Distribution	Reservoir	Transmission	Aerosol	Incubation (days)	Ribavirin effective	Vaccine	Mortality (%)
Filoviridae									
Ebola	Ebola Hemorrhagic fever	West Africa	Unknown	Person to person, nosocomial	Yes, in animal models	2–21	No, but Ig may be helpful	No, but promising developments	50–90
Marburg	Marburg virus disease	East Africa/Sub-Saharan	Unknown	Person to person, nosocomial	?	3–9	No	No	~25
Arenaviridae									
Lassa	Lassa fever	West Africa	Rodents, *mastomys huberti and erythroleucus*	Aerosol, Person to person, sexual	Yes	5–16	Yes	No	15–25
Junin	Argentinian Hemorrhagic fever	Argentina	Rodent, *Calomys musculinis*	Aerosol, Person to person, sexual	Yes	7–14	Probably, Convalescent plasma	Yes	15–30
Machupo	Bolivian Hemorrhagic fever	Bolivia	Rodent, *Calomys callosus*	Aerosol, Person to person, sexual	Yes	7–14	Probably, Convalescent plasma	No	18

Family/virus	Disease	Normal Distribution	Reservoir	Transmission	Aerosol	Incubation (days)	Ribavirin effective	Vaccine	Mortality (%)
Guanarito	Venezuelan Hemorrhagic fever	Venezuela	Rodent, *Zygodontomys brevicauda*	Suspected to be by aerosol	Yes	? but similar to Junin	Unknown but ribavirin suggested	No	~25
Sabia	Brazilian Hemorrhagic fever	Brazil	Unknown	?	Unknown	? but similar to Junin	Unknown but ribavirin suggested	No	?
Bunyaviridae									
Rift Valley	Rift Valley Fever	Sub-Saharan Africa	Sheep/domestic animals	Mosquito bite, aerosol, animal blood	Yes	2–6	Unknown	Yes, experimental	1
Crimean–Congo	Crimean–Congo Hemorrhagic Fever	Africa, Asia, Europe	Domestic animal, hedgehogs, rabbits	Ticks (*Hyalomma marginatum*)	Yes	2–12	Yes	No	50
Hantavirus Hantaan group	Hemorrhagic fever with renal syndrome	Worldwide	Rodents	Aerosolized mouse droppings	Yes	5–42	Yes	No	5
Hantavirus Sin Nombre group	Pulmonary syndrome	North, South America	Rodents (*Peromyscus maniculatus* and *Sigmondon hispidus*)	Aerosolized mouse excrement	Yes	8–21	No	No	~47–63

Family/virus	Disease	Normal Distribution	Reservoir	Transmission	Aerosol	Incubation (days)	Ribavirin effective	Vaccine	Mortality (%)
Flaviviridae									
Kyasanur Forest virus	Kyasanur Forest disease	Western India	Monkeys, rodents, insectivores, domestic live stock	Ixodid ticks Haemaphysalis spinigera	Not known	3–8	Unknown	Yes	
Osmk Hemorrhagic Fever virus	Osmk hemorrhagic fever	Western Siberia	Rodents, muskrats, water voles (Arvicola terrestris)	Ticks Dermacentor reticulatus	Not known	?	Unknown	No, but cross protection from Tick-borne encephalitis virus	0.5–3
Alphaviruses									
Eastern Equine Encephalitis virus	Encephalitis	North, Central and South America, Caribbean	Mosquito, bird, mammal	Mosquito, Aerosol	Yes	3–10	No	Yes but experimental	24–56 depending one age
Western Equine Encephalitis virus	Encephalitis	North America, Brazil, Argentina, Uruguay	Mosquito, bird, hare	Mosquito, Aerosol	Yes	2–10	No	Yes but experimental	4

9.2
Filoviridae

9.2.1
Virology

This family contains both Ebola and Marburg virus. Filoviruses have a filamentous morphology (Fig. 9.1), a negative sense RNA genome (19,000 bases), and are enveloped. The RNA produces monocistronic messages and encodes for several important proteins. The highly glycosylated, homotrimer, transmembrane spike protein varies in size between the two viruses, with Ebola and Marburg measuring 125 kDA and 170 kDA, respectively [2]. Much has been written about the importance of the Ebola and Marburg glycoprotein with regard to virus entry, molecular virology, infectivity, and pathogenesis and possible vaccine target [3–16]. The Ebola GP is unique in that it is synthesized from a precursor GP_0 that is post-translationally cleaved into two subunits, GP_1 and GP_2. These two subunits are linked by disulfide bonds to form a heterodimer. Homotrimers of this complex form the virion spikes [17, 18]. There is also a soluble, non-structural secreted glycoprotein (sGP), formed from unedited transcripts; it has been postulated that this has a role in establishment and maintenance of infection in the natural reservoir, possibly as a result of some type of immunosuppressive effect [19]. Other proteins are the VP40 (matrix), VP30 (minor nucleocapsid), VP24 (membrane protein), nucleoprotein (major nucleocapsid), and polymerase (RNA-dependent RNA polymerase). Ebola virus has four known genetically heterogeneous subtypes – Zaire, Sudan, Ivory Coast, and Reston. Within Ebola subtypes, there is little genetic heterogeneity, suggesting a stable relationship with the reservoir host [19]. Marburg virus has no known subtypes.

9.2.2
Epidemiology

There have been several recorded outbreaks of Ebola and Marburg virus. The first outbreak occurred in 1967 among European laboratory workers handling African

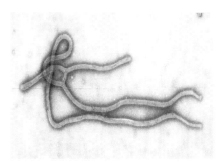

Fig. 9.1
Ebola virus (electron microscopy).

green monkey tissue. The outbreak lead to the discovery of Marburg virus. African green monkeys were brought from Uganda and harvested for tissue. There were 31 cases of human infection with 25 primary and six secondary cases. Seven fatalities were associated with this initial outbreak [20]. A second outbreak occurred in 1980 when a French engineer working in western Kenya, and his physician, contracted Marburg disease [21]. The most recent outbreak of Marburg occurred in the Uige Province of Northern Angola [22].

The first recognized outbreaks of Ebola were recorded in 1976 and were seen in northern Zaire and in southern Sudan [23, 24]. Ebola was shown to be antigenically different from Marburg. It was also found that Ebola Zaire was different from Ebola Sudan. In 1994 a new Ebola subtype was identified in Côte-dIvoire by an ethologist who performed a necropsy on an infected chimpanzee [25]. In November, 1998 a Ebola subtype with little pathogenicity for humans was recognized when an animal care worker noticed a high mortality rate in cynomolgus monkeys in an animal-care facility. An investigation of the outbreak revealed Ebola Reston and a subsequent epidemiological investigation revealed four cases of asymptomatic human infection [26, 27].

There have been several epidemiological investigations after each individual case or outbreak of Ebola or Marburg virus. The viral reservoirs for these viruses have not been determined nor is it known how they enter the human population. The viruses are, however, commonly passaged in animals or mammalian cell lines, suggesting their original reservoir may not be an insect. Interestingly, Ebola virus may replicate to high titer in fruit and insectivorous bats without antibody production. Virus may be recovered from the feces of these animals after experimental infection [28].

9.2.3
Clinical Manifestations and Disease

There seems to be little difference between the clinical pictures of Ebola and Marburg hemorrhagic fevers. Filovirus hemorrhagic fevers have an incubation period of 5 to 10 days with a range of 2–19 days. They begin suddenly, with initial onset of fever, chills, myalgia, and severe frontal headache. Accompanying the fever may be nausea, vomiting, abdominal pain, chest pain, photophobia, jaundice, pancreatitis, and central nervous system involvement. This may be manifested as delirium or coma [2, 19, 21, 23, 24, 29–31]. Hiccups have been associated with infection in 5–18% of patients [32]. Patients deteriorate quickly and progress to severe nausea, vomiting, hematemesis, and melena. There may be a purple–red maculopapular rash in 50% of Zaire Ebola virus-infected subjects around day 5, although a lower incidence of rash was observed in a more recent outbreak [31] (Table 9.3). A ghostlike facies has also been described [33].

Tab. 9.3

Signs and symptoms of fatal and non-fatal Ebola virus
infection [31].

Characteristic	Fatal cases (%, $n = 84$)	Non-fatal cases (%, $n = 19$)
Asthenia	85	95
Diarrhea	86	84
Nausea and vomiting	73	68
Abdominal pain	62	68
Headache	52	74
Sore throat, odynophagia, or dysphagia	56	58
Arthralgia or myalgia	50	79
Lumbar pain	12	26
Chest pain	10	5
Fever	93	95
Conjunctival injection	42	47
Tachypnea	31	0
Hiccups	17	5
Rash	14	16
Anuria	7	0
Gums bleeding	15	0
Hematemesis	13	0
Melena	8	16
Hematuria	7	16
Puncture site bleeding	8	5
Bloody stools	7	5
Petichiae	8	0

Spontaneous abortion occurs during acute infection with filoviruses [23, 24]. During the Ebola virus outbreak in Zaire in 1977, nine children from infected women died within 19 days of birth [23]. Handling of fetal tissues has been associated with transmission [23, 32].

Laboratory abnormalities are common. Leukopenia and thromobocytopenia are present. Leukocytosis may ensue later in the disease. In animal models, abnormalities of the coagulation system are present from the beginning of infection [34]. Liver enzyme elevations are present along with proteinuria [19, 30].

Death ensues secondary to shock 6 to 9 days after the onset of clinical disease. Autopsy results have shown hemorrhage into the skin, mucous membranes, visceral organs, and gastrointestinal tract. Microscopic changes reveal necrosis in the liver, kidneys, lymphatic organs, and testes. The liver is usually prominently involved (Fig. 9.2). Lung involvement occurs in the form of interstitial pneumonitis. There is currently no treatment for either of these viruses. In subjects that survive, convalescence may be prolonged. There may be multiple complications

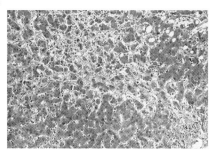

Fig. 9.2
Liver necrosis in Ebola Zaire disease.

including orchitis, transverse myelitis, uveitis, recurrent hepatitis, hearing loss or tinnitus, and arthralgias [31, 35, 36]. Treatment is supportive and includes maintaining fluid balance and replacement of coagulation factors and platelets.

In the 1976 Zaire and Sudan outbreaks there was 550 cases of Ebola and more than 430 deaths. The death rate for Ebola disease has varied depending on the subtype. Ebola Reston did not cause any mortality in humans. The mortality rate from infection with the other three subtypes has ranged from 53 to 92 % [19]. The reason for this variation is not known. Marburg virus mortality has been lower but is still substantial with a range of 23 to 33 % [19]. Asymptomatic or minimally symptomatic infections have, however, been suggested by serosurveys after Ebola virus outbreaks. A study using an ELISA method after a 1995 Zaire Ebola outbreak suggested there were other outcomes besides death or severe disease after infection [37, 38]. Serum collected from close contacts of subjects after an Ebola outbreak in Gabon contained high levels of both circulating cytokines and Ebola-specific antibody, further suggesting infection may occur without overt disease [39].

9.2.4
Transmission

The method of transmission for both viruses is not known. Possible mechanisms of transmission include direct contact, droplet spread, fomites and body fluids, and airborne transmission via small-particle aerosols. It seems that isolating patients and adhering to standard infection control and barrier precautions can interrupt transmission. Retrospective analysis of the Ebola Zaire epidemic of 1995 suggested that people who came into contact with body fluids including vomit, blood, feces, urine, and sweat were at higher risk of secondary infection. In primate infection models, aerosolized Ebola virus is highly infectious [34]. It has been suggested as little as 30 plaque-forming units (PFU) of Marburg virus are enough to initiate infection of the host. This was determined by aerosol infection of green monkeys [40]. A characteristic of all viruses is their inability to survive outside the host. It has been found in a laboratory chamber that Marburg virus, when aerosolized, loses approximately 10 % infectivity per minute, although simple addition of a common laboratory chemical can reduce this to 1 % per minute [41, 42].

Both Ebola virus and Marburg virus are CDC category A Bioterrorism agents [43]. As such, these agents are handled in a BSL-4 laboratory. Ideally, Filovirus-infected patients should be taken care of by personnel well versed in barrier precautions. Both aerosol and contact precautions should be enforced, including negative pressure rooms, impermeable gowns, gloves, face shields, and respirators. In a situation in which the health care system may be overwhelmed, contact precautions should still be enforced. Fatalities should either be cremated or promptly buried.

9.2.5
Pathogenesis

A more detailed picture is emerging about the pathogenesis of Ebola virus disease. Ebola can replicate in a wide variety of cells. It seems that macrophages and dendritic cells are the initial sites of Ebola virus infection and seem to be the source for pro-inflammatory cytokines [44–46]. Virus is transported to regional lymph nodes by macrophages and dendritic cells, where further replication occurs, followed by dissemination through lymphatic channels to the spleen, liver, lymph nodes, and other tissues. Viral replication results in extensive necrosis. Recent studies have suggested that fatal illness is associated with increasing titers of virus in the bloodstream [47, 48]. *In-vitro* studies have suggested that both bound and soluble glycoprotein may be directly cytotoxic [49]. It has been suggested that the glycoprotein from Zaire Ebola virus may preferentially bind to endothelial cells, killing them, resulting in increased vascular permeability [8]. The role of either GP or sGP in the pathogenesis of disease remains unanswered.

In animal models, lymphocyte depletion is common in many different organs [50]. This seems to be because of massive apoptosis of lymphocytes [46, 50]. Several proapoptotic genes seem to be upregulated by day 3 after infection. TNFα-related apoptosis-inducing ligand (TRAIL), FAS-associated death domain, and FAS/CD95 mRNA levels have been found to be elevated [46]. Indirect evidence for the activation of apoptotic pathways in humans have been suggested, including FAS mediated and caspase-independent apoptosis in people infected with Ebola [50, 51]. Apoptosis occurs late in infection and may affect the adoptive immune response [52].

Filovirus infection of dendritic cells may result in interference in both adaptive and innate immune responses. Filovirus-infected dendritic cells are inhibited from releasing interferon α, interleukin 1β, and TNFα, although both macrophage inflammatory protein (MIP-1α) and monocyte chemotactic protein 1 (MCP-1) are secreted by these cells. The expression of these cytokines requires the action of nuclear factor ϰβ (NF-ϰβ). It seems there is a block by Zaire Ebola virus, not of the translocation of NF-ϰβ to the nucleus, which would interfere with expression of all the preciously mentioned cytokines, but some place downstream that enables transcription of only MIP1α and MCP-1 [53, 54]. VP35 may suppress innate responses by suppressing production of interferon production. In *in-vitro* experiments VP35 has been found to be able to replace the influenza NS1 protein in its ability to suppress interferon production [55]. It was later shown that VP35 impedes the phosphorylation and nuclear translocation of interferon response factor 3, inhibiting the transcription of interferon related genes [34, 56]. It seems VP35 might also suppress α interferon as shown in a chimeric alphavirus model [57].

One of the most important features of Ebola or Marburg virus disease is hemorrhagic fever and shock characterized by disseminated intravascular coagulation and increased vascular permeability [14, 34, 45, 58–61]. Macrophage and dendritic cell infection with Ebola virus leads to tissue factor (TF) generation, fibrin deposition, and microthrombosis [45, 60]. Primate models have suggested that endothelial cells in lymphoid organs may be the initial site of Ebola replication and not widespread endothelial involvement, as was first believed [45]. Within 24 h of infection, laboratory coagulation abnormalities can be observed, as measured by an increase in D-dimers. This laboratory finding precedes clinical signs and seems to be a consequence of the expression of tissue factor on virus-infected macrophages [59]. Microparticles containing tissue factor are released from the macrophages into the general circulation, contributing to the systemic coagulopathy. Forty-eight hours after infection, protein C levels are also reduced. Platelet counts fall on days 3–6, indicative of ongoing coagulopathy [60]. Interestingly, the course of the disease may be affected by giving a TF-specific hookworm inhibitor nematode anticoagulant peptide c2 (NAPc2) to primates infected with Ebola. By giving NAPc2 within 24 h of infection, and then daily for 8 or 14 days mortality was reduced by 33%. This treatment also reduced D-dimer levels and suppressed the levels of inflammatory cytokines including IL-6 [58]. Ebola virus titers were also reduced in the survivors, suggesting that intervention in TF production may either enhance innate immunity or limit viral replication and spread.

Both innate and specific immunity seem important in recovery from Ebola virus disease. Subjects who recover have an early appearance of pro-inflammatory cytokines [51, 62, 63]. An IgG-mediated response has been associated with survival [39, 64]. In a murine model, Ebola-like particles composed of the GP and matrix viral protein VP40 with a lipid envelope were used to immunize mice. This produced a protective IgG subclass response, CD4+, and CD8+ interferon gamma-producing cells. It was found that both humoral and cytotoxic CD8+ cells were important for protection from lethal challenge [65].

9.2.6
Diagnosis

Differential diagnosis depends heavily on travel history or possible exposure to either Marburg or Ebola virus and the time of presentation by the patient. Differential diagnosis is very broad if the patient presents early with fever, headache, and diarrhea and with a travel history. If the subject has bleeding or manifestations of disseminated intravascular coagulation, the diagnosis would be narrower but would include other hemorrhagic fever viruses and early hemorrhagic smallpox [19]. Other illnesses to consider include malaria, tick-borne typhus, leptospi-rosis, plague, and borelliosis. If intentional release of the virus occurred in a major urban center, initial symptoms might resemble an enterovirus or other viral infection.

Diagnosis of Ebola or Marburg infection has been accomplished by polymerase chain reaction [38, 47]. Virus can be isolated from the bloodstream during acute infection and seroconversion occurs relatively early on days 8–12. ELISA assays may be used in early convalescence for detecting IgM antibodies [47, 66]

9.3
Arenaviridae

9.3.1
Virology

The Arenaviridae are divided into two major antigenic and phylogenetic groups designated Old and New World. The Old World group includes lymphocytic choriomengitis virus (LCM), Lassa fever virus, and other, related, viruses whereas the Tacribe complex or New World group includes the American viruses such as Junin, Machupo, Guarnarito, and Sabia viruses. The Arenaviridae viruses are limited in their geographic niche by their rodent host. The Old World viruses have evolved in a close relationship with rodents of the family Muruide, Sub-family Murinae whereas the New World viruses are associated with family Muruide, sub-family Sigmondontinae. The New World viruses are further divided into Clades A, B, and C. The Tacribe virus seems to be the only one that does not have a rodent host; it was isolated from bats [67–79]

The arenavirus family is characterized by having a single stranded, ambisense RNA genome. The genome is divided into a two single stranded RNA molecules

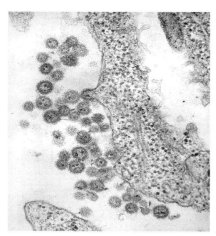

Fig. 9.3
Electron micrograph of Machupo virus, the agent of Bolivian hemorrhagic fever. Note granular appearance of virions that gives Arenaviridae its name.

designated S (small) and L (large) in the molar ratio 2:1, respectively. The S and L segments are approximately 3.3 to 3.5 and 7.2 kDa, respectively [80]. The virions are enveloped and have an average diameter of approximately 130 nm. When visualized under electron microscopy, the virions can be seen to have granules, which have been shown to be host ribosomes. It is this appearance that has given the family its name from Latin, *arenosos*, or sandy (Fig. 9.3).

9.3.2
Epidemiology

The distribution of arenaviruses is based on their rodent host which is usually one species, although exceptions have been noted [80]. The dynamics between virus and rodent are one of chronic infection enabling excretion of the virus in rodent body fluids including urine and feces. Transmission from rodent to rodent is vertical and horizontal, enabling persistent infection within the population. It is the juxtaposition of man and rodent that enables entry of the virus into the human population.

Lassa fever virus cases have mainly come from Nigeria, Liberia, Sierra Leone, and Guinea. Most outbreaks occur during the dry season from January until April, although cases are reported throughout the year. Serological studies indicate that infection is common and in some villages in Sierra Leone is as high as 10 to 20% [81–83]. Infection is usually subclinical and it has been estimated there are 20 subclinical or minimally apparent infections for each overt case of disease [82].

Argentine hemorrhagic fever occurs in a restricted, but expanding, geographical region north of Buenos Aries province. Infection is most common in February to May when the corn harvest occurs. Bolivian hemorrhagic fever is found in north-eastern Brazil. Most infections are acquired as a result of household by contact with infected rodent droppings. The incidence is greatest in April to July during the late rainy and early dry seasons [83].

9.3.3
Clinical Manifestations and Disease

Incubation periods vary from virus to virus but range from 5 to 16 days and from to 7 to 14 days for Lassa fever and the South American hemorrhagic fevers, respectively. Lassa fever virus illness begins gradually, with fever, malaise, asthenia, and gastrointestinal symptoms. Chest pain, cough, abdominal pain, diarrhea, vomiting, and sore throat may occur [82–87]. One study suggested the triad of proteinuria, pharyngitis, and retrosternal pain has a predictive value of 80% and a specificity of 89% [84]. As the disease progresses there may be bleeding, conjunctival injection, rales, and facial edema. Hepatic enzyme abnormalities may be seen and one study correlated serum aspartate aminotransferase levels greater than 150 IU L^{-1} with an increased mortality of 50% and a need for ribavirin therapy [88].

Neurological findings are less common than in Argentine hemorrhagic fever but may occur, leading to encephalitis, encephalopathy, deafness, and cerebellar syndromes [80, 84, 89, 90]. Death is usually seen in the second or third week of illness and is characterized by hypotension, reduced urinary output, facial and pulmonary edema, and capillary leak syndrome. Minor hemorrhages may be present. In patients who survive, deafness was observed in one third of hospitalized survivors [89]. Other long-term complications of the disease include uveitis and orchitis [83].

The incubation period of both Argentine and Bolivian hemorrhagic fever is between 7 and 14 days. They are similar in clinical presentation and may begin insidiously with onset of fever, myalgia, and malaise. There may be additional clinical manifestations including vomiting, diarrhea, dizziness, retro-orbital pain, and photophobia [91–93]. A rash may also be present in the form of petechiae or small vesicles on the palate and the fauces, and on the skin in the axilla [83]. The most important complications of the disease are hemorrhagic or neurological. These presentations may occur together or separately [94]. Neurological signs are common. The disease may progress to confusion, cerebellar tremor, gait abnormalities, hyporeflexia, and tongue tremors. If the patient experiences seizures and coma, the prognosis is poor. Alternatively, the patient may shows signs of capillary leak syndrome, decreasing urinary output, and early bleeding manifestations. In severe cases, the patient may progress to hemorrhage from the gastrointestinal tract and have coma, seizures, and shock. The mortality rate ranges from 15 to 30%. Convalescence may take several weeks but is usually without sequelae. Alopecia, hypotension lasting 1 to 2 weeks, and Beau lines in nails are seen. Intrauterine infection is common in arenavirus infections. Lassa fever virus infections of pregnant women are usually accompanied by spontaneous abortion and a high maternal death rate [95].

As with other hemorrhagic fever viruses, supportive care is important but several of these viruses may respond to antiviral therapy. Argentine hemorrhagic fever can be treated with convalescent serum if given within the first 8 days of the disease; this results in a decrease in the case-to-fatality ratio from ~20% to 1% [96–98]. A late neurological syndrome has been characterized in patients treated with immune plasma; this is characterized by cerebellar signs, fever, headache, and dizziness that may last for several days then resolving without permanent sequelae [96, 99]. The logistics of locating, producing, storing, and delivering human plasma is difficult. Ribavirin has been used in the treatment of arenavirus infections. Ribavirin has been given intravenously to subjects at high risk of death and mortality was reduced from 50 to 5% if given 7 days after the onset of disease, although mortality was reduced at all times after presentation [100]. Patients were given a 30 mg kg^{-1} loading dose followed by 15 mg kg^{-1} every 6 h for fours days, then reduced to 7.5 mg kg^{-1} every 8 h for six more days. The only side effect was a reversible anemia that did not require transfusion. Ribavirin seems to have activity against other arenaviruses and should be considered as therapy in any arenavirus infection. [92, 94, 101–108]. A major success in arenavirus prevention has been the development of a live attenuated vaccine for Junin virus [109]. Since the introduction of this vaccine, it has drastically reduced the incidence of Argentine hemorrhagic fever and may provide some protection against Machupo virus [83].

9.3.4
Transmission

It seems that most infections by arenaviruses are via aerosol transmission. Junin virus is common is in workers using manual or mechanical methods for corn harvesting. It is most likely the close proximity of people to rodents and their aerosolized excrement is a plausible means of transmission. Multiple laboratory infections with arenaviruses have also demonstrated the high infectivity of these viruses when aerosolized. Aerosol infection within hospitals has also been documented [81, 110]. When possible, patients should be placed in a negative pressure room and aerosol and contact barrier nursing should be used. Decontamination should be with sodium hypochlorite.

Arenaviruses are potential bioterrorism agents. The arenaviruses discussed are all listed as Category A bioterrorism pathogens. It seems they require a low inoculum to infect and cause disease with the 50% lethal dose for Junin and Lassa fever being 50 and 500 PFU, respectively [42]. In addition, they can be grown easily is tissue culture. They are known to initiate infection and disease through the respiratory tract [94].

9.3.5
Diagnosis

Diagnosis of Lassa fever virus infection is usually accomplished by ELISA [83, 88]. Virus may be isolated from all arenavirus hemorrhagic fever cases during acute infection. Lassa fever virus may remain persistent in the blood longer than other arenaviruses [88]. Junin virus could be isolated from mononuclear cells by co-cultivation on Vero cells [111] whereas Machupo virus may be more difficult to isolate. Reverse transcription then PCR has been accomplished for all arenavirus infections [83, 112].

9.4
Bunyaviridae

9.4.1
Virology

The Bunyaviridae encompass a broad range of viruses and some are important causes of human disease. The Bunyaviridae are divided into five genera including

Bunyavirus, Nairovirus, Phlebovirus, Hataan, Tospovirus, and Phlebovirus. There are three viruses of interest that should be mentioned as possible bioterrorism weapons including the Hantaviruses, Crimean-Congo hemorrhagic fever virus (CCHF), and Rift Valley fever virus (RVF). The Bunyaviridae are enveloped viruses approximately 100 nm in diameter with glycoprotein spikes and a single stranded, negative polarity RNA genome. The envelope surrounds a core that has three nucleocapsids, each containing a different length of RNA. This RNA is designated short (S), medium (M), and long (L). There are three structural proteins produced including two glycoproteins and one nucleoprotein [113].

9.4.2
Epidemiology

The distribution of the Bunyaviridae is worldwide. The Hantaan group represented by the Hantaan and Seoul viruses are found in Asia and Dobrava is found in Europe. The Sin Nombre group is found in North and South America. Rift Valley fever virus is seen in Sub-Saharan Africa and is more prevalent during the rainy season. In 1977, it was introduced into Egypt, causing widespread disease in animals and humans [114–120]. Congo–Crimean hemorrhagic fever virus is found in Africa, Southeast Europe, the Middle East, and Asia.

9.4.3
Clinical Manifestations and Disease

Hantaviruses may cause two different syndromes. In May 1993, there was an outbreak of unexplained deaths of young adults in New Mexico, Arizona, Colorado, and Utah (Four Corners) [121–123]. These patients died of a pulmonary syndrome consistent with ARDS, known as Hantaan pulmonary syndrome (HPS), and was eventually attributed to a virus named Sin Nombre (Fig. 9.4). HPS begins as a febrile syndrome characterized by myalgia, cough, nausea, and vomiting. These symptoms may be accompanied by dizziness, arthralgias, and, late in disease, shortness of

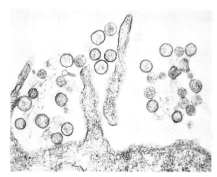

Fig. 9.4
Electron micrograph of Sin Nombre virus.

breath. Rhinorrhea and sore throat are rare symptoms [124–126]. Physical examination may be remarkable for tachypnea, hypotension, tachycardia, and rales on lung examination. The chest radiograph shows bilateral interstitial infiltrates that may progress rapidly. Laboratory findings may show thrombocytopenia, atypical lymphocytes, and a left shift. Liver enzymes may be elevated [126–128].

The other clinical presentation seen in Hantavirus infections is hemorrhagic fever with renal syndrome (HFRS). The incubation period may vary from 5 to 42 days averaging approximately 2 weeks. In the toxic stage, patients may present with complaints similar to those of HPS but without the pulmonary signs. These include headache, lower back pain, dizziness, abdominal pain, and blurred vision [129–134]. There may be conjunctival injection and petechiae on the trunk and soft palate. The laboratory profile may be the same as in HPS. After the patient has become afebrile (4 to 7 days), shock may ensue with a hemorrhagic picture. This stage of the disease is when most deaths occur. Patients who survive will pass through an oliguric phase then a non-oliguric phase. Complete recovery may take months.

No vaccines are available for any of the Hantaviruses. Most care is supportive including strict attention to fluids and the treatment of any complications. The role of ribavirin in the treatment of each of these syndromes is variable. In subjects with HPS, ribavirin has not been found to be effective but further research is ongoing [134]. There has been one prospective, placebo-controlled trial with ribavirin for HFRS in the Peoples Republic of China [135]. The study enrolled 242 serologically confirmed cases of HFRS, excluding subjects with more than 6 days of symptoms, shock refractory to fluid replacement, coma, or renal failure. Overall, there was a sevenfold decrease in mortality in the treated group compared with the untreated group. In addition, therapy reduced both hemorrhage and oligouria.

Rift Valley fever has an incubation period of 2–6 days with most patients developing fever, myalgias, and malaise [118, 136–143]. Most subjects recover without incident but approximately 1% of subjects will progress to fulminant disease including hepatitis, jaundice, and hemorrhage [144]. There may be a prolonged convalescent period. Another 1% of patients may progress to CNS involvement including encephalitis and ocular complications. The disease expresses itself in a biphasic manner with resolution of the initial febrile syndrome followed in several days to weeks by signs of encephalitis. This may include headaches, coma, hallucinations, focal neurological signs leading to death. Ocular findings include decreased visual acuity, retinitis with exudates, and uveitis [145, 146].

A formalin inactivated vaccine has been shown to be effective in trials [147–149]. This vaccine is not licensed. There is currently no recommendation for the use of ribavirin for treatment of RVF.

Crimean–Congo hemorrhagic fever is a disease first described in Russia in the 12th century [150]. The clinical disease begins as an acute febrile illness with fever, chills, headache, and myalgias. This may be followed by nausea, vomiting, and abdominal pain [134, 151, 152]. There may be flushing of the face and conjunctival injection with hepatomegaly present in 50 % of cases. Patients seem to be getting better but after 3–6 days lapse into the hemorrhagic part of the disease. This includes a petichial rash over the trunk and limbs extending to the whole body, ecchymoses, and bleeding from the gums, nose, uterus, lungs, and gastrointestinal tract. There may be CNS involvement in 25 % of subjects including nuchal rigidity, excitation, depression, and coma. CNS involvement portends a poor prognosis [152]. Laboratory findings include thrombocytopenia, leukopenia, increased transaminase levels, proteinuria, and increased creatinine.

There is no vaccine for CCHF. Ribavirin is the drug of choice although experience is limited [101, 153–160].

9.4.4
Transmission

The Hantaan and Sin Nombre virus groups are spread by aerosolization of rodent excrement and urine. This is an exception to many of the viruses of the Bunyaviridae, which usually have an insect vector. As with the arenaviruses, rodent infection with virus is a chronic, persistent infection and the virus may be excreted for weeks to months. Human disease corresponds to an increase in the rodent population [161–167]. RVF in Africa may be transmitted in the blood of infected animals and by mosquito bites. The virus is maintained by transovarial passage of virus in mosquito eggs that may hatch when they come into contact with heavy rains. CCHF virus infects humans most commonly after the bite of an adult *Hyalomma* tick, although nosocomial transmission has also been documented [168]. All of the viruses described above have been implicated in aerosol infections naturally, in the laboratory, or in the hospital. [153, 168–173] As such, they are potential bioterrorism weapons. RVF disease requires a small inoculum to initiate disease [42]. The Hantaviruses are difficult to culture, making them less attractive for scaling up to an amount for large-scale dispersal.

9.4.5
Diagnosis

Almost all patients presenting with Hantavirus infections have both IgM and IgG as detected by an ELISA [134]. As mentioned above, viral isolation may be difficult in either tissue culture or animals, but reverse transcription PCR has been successfully used [128]. Both RVF and CCHF may be cultured from subjects during the acute disease phase. Both ELISA and reverse transcription PCR assay have been used for diagnosis.

9.5
Flaviviridae and Other Viruses

The prototype viruses for the Flaviviride family are yellow fever and dengue virus. These viruses do not seem to be significant threats as bioterrorism weapons. Yellow fever virus has a mortality rate of 30% but an effective vaccine is available and could probably be mass produced in an emergency. Dengue fever virus is not a serious disease unless there is infection with a second serotype in which the manifestations of Dengue hemorrhagic fever with shock become evident. As such, neither would be as attractive as other members of this family unless they were engineered to enhance virulence or to overcome vaccine protection. There are other viruses in this family that could be used as bioterrorism weapons.

9.5.1
Kyasanur Forest Disease

Kyasanur Forest disease (KFD) was first reported from the Karnataka State in India. It was found to be a new member of the tick-borne flaviviruses. KFD was transmitted by ixodid ticks while maintaining a reservoir in monkeys, insectivores, and forest rodents [174–176]. Sporadic cases of KFD occur during the dry season and occur in workers clearing forests for farmland. The incubation period of the disease is between three to eight days. Symptoms begin abruptly with fever, chills, vomiting, headache, myalgia, cough, gastrointestinal symptoms, photophobia, and conjunctival suffusion. On examination, petechiae, hepatosplenomegaly, lymphadenopathy, facial and conjunctival hyperemia, and bradycardia may be observed [177]. Virus may be recovered from the blood in the first week of illness.

Hemorrhagic complications may ensue, including bleeding from the nose, gums, gastrointestinal tract, and hemorrhagic pulmonary edema. KFD is a biphasic illness with the first phase lasting 6 to 11 days; this is followed by an afebrile period of 9 to 21 days. The second phase of the disease occurs in 15 to 50% of patients and is characterized by meningoencephalitis. Five to ten percent of cases are fatal. A vaccine is made in chicken embryos and is available locally in response to outbreaks [178–180].

9.5.2
Omsk Hemorrhagic Fever Virus

Less is known about Omsk hemorrhagic fever virus (OHF). It was first isolated in 1947 in Siberia and is found in the forests of Western Siberia in the Omsk, Novosibirsk, Kurgan, and Tjumen regions. Dermacentor ticks transmit the virus and small animals, including water voles, seem to be the reservoir [181]. The disease became more prevalent after the introduction of muskrats to stimulate a fur industry [182]. Approximately 1500 human cases were reported between 1945 and 1958 among trappers, family members, and laboratory workers. The virus seems to be acquired directly from the animals or transmitted by ticks. The disease resembles KFD but with more frequent neurological sequelae including hearing loss and neuropsychiatric symptoms. Hair loss is common. The case fatality rate is between 0.5 and 3%.

9.6
Alphaviruses

The alphaviruses are not hemorrhagic fever viruses but have the potential to be biological weapons. These viruses are enveloped, positive-stranded RNA viruses in the family Togaviridae. They are transmitted by mosquitoes to other animals with man being an incidental host. They have many of the characteristics of a bioweapon including ease of cultivation, person-to-person spread by aerosol, relatively small inoculum to initiate disease, and potential for genetic manipulation. Another favorable characteristic is that most human infections are symptomatic. It has been modeled that 50 kg Venezuelan equine encephalitis virus released downwind from a city of 500,000 people would spread 1 km from the point of release and cause disease in 35,000 people, with 400 deaths [42, 183]. If released upwind, it would be distributed in a 20-km area and would infect 220,000 people, with 95,000 deaths. In terms of the number of subjects infected or the lethality of equal amounts, however, hemorrhagic fever viruses do not compare with anthrax spores.

Eastern equine encephalitis (EEE) virus is found only in the Western Hemisphere, including all of the Americas. Human cases occur sporadically [184]. The incubation period has not been defined but it likely to be 3–10 days. There is a 5–10 day prodrome of headache, nausea, vomiting, diarrhea, and chills. These initial symptoms are then followed by CNS disease including seizures, mental confusion, cranial nerve palsies, spastic paralysis, bilateral papilledema, and coma. Laboratory abnormalities include elevated white counts with a left shift, hyponatremia, and elevated CSF cells (lymphocytes) and protein. Mortality varies with patient age, from 30% (children <20 years) to 56% (adults >60 years); it is 24% in middle age [184–186]. Permanent sequelae included behavioral changes, retardation, seizure disorders, and paralysis. Western equine encephalitis (WEE) virus also has a short incubation period of 2–10 days. It has a similar presentation, with a prodromal period of headache, dizziness, fever, myalgia, and malaise. Neurological complications are not as prominent and the overall mortality is 4% [187]. Venezuelan equine encephalitis (VEE) virus has the shortest incubation period, which ranges from 12 h to 2–3 days [188, 189]. The usual presentation of VEE is a febrile illness that progresses to encephalitis in 1 and 4% of adults and children, respectively.

Summary

Several families of viruses described in this chapter could be used for bioterrorism. The hemorrhagic fever viruses are, because of the presentation of their clinical disease syndromes, frightening to clinicians, governments, and the general public. A cluster of cases of Ebola or Marburg hemorrhagic fever in a major urban center would be a major crisis and require the full cooperation of health-care workers, public health, state and federal government, and the media to manage. Further work is needed for research and development of vaccines, antivirals, and state and federal responses to a bioterrorist event.

References

1 Borio, L., Inglesby, T., Peters, C. J., et al. **2002** Hemorrhagic fever viruses as biological weapons: medical and public health management.[see comment]. *JAMA*; 287(18):2391–2405.

2 Sanchez, A., Khan, A. S., Zaki, S. **2001** Filoviridaie: Marburg and Ebola viruses. In: Knipe D, Howley P, editors. *Virology*. 4th ed. Philadelphia: Lippincott, Williams, and Wilkins; p. 1279–1304.

3 Volchkov, V.E., Blinov, V.M. and Netesov, S. V. **1992** The envelope glycoprotein of Ebola virus contains an immunosuppressive-like domain similar to oncogenic retroviruses. *FEBS Letters*; 305(3):181–184.

4 Becker, Y. **1995** Retrovirus and filovirus "immunosuppressive motif" and the evolution of virus pathogenicity in HIV-1, HIV-2, and Ebola viruses. *Virus Genes*; 11(2/3):191–195.

5 Feldmann, H., Volchkov, V.E., Volchkova, V.A., et al. **1999** The glycoproteins of Marburg and Ebola virus and their potential roles in pathogenesis. *Archives of Virology – Supplementum*; 15:159–169.

6 Chepurnov, A. A., Tuzova, M. N., Ternovoy, V. A., et al. **1999** Suppressive effect of Ebola virus on T cell proliferation in vitro is provided by a 125-kDa GP viral protein. *Immunology Letters*; 68(2/3):257–261.

7 Wool-Lewis. R. J. and Bates. P. **1999** Endoproteolytic processing of the ebola virus envelope glycoprotein: cleavage is not required for function. *Journal of Virology*; 73(2):1419–1426.

8 Yang, Z. Y., Duckers, H. J., Sullivan, N. J., et al. **2000** Identification of the Ebola virus glycoprotein as the main viral determinant of vascular cell cytotoxicity and injury. *Nature Medicine*; 6(8):886–889.

9 Chan, S. Y., Speck, R. F., Ma, M. C., et al. **2000** Distinct mechanisms of entry by envelope glycoproteins of Marburg and Ebola (Zaire) viruses. *Journal of Virology*; 74(10):4933–4937.

10 Takada, A. and Kawaoka, Y. **2001**The pathogenesis of Ebola hemorrhagic fever. *Trends in Microbiology*; 9(10):506–511.

11 Warfield, K. L., Bosio, C. M., Welcher, B. C., et al. **2003** Ebola virus-like particles protect from lethal Ebola virus infection. *Proceedings of the National Academy of Sciences of the United States of America*; 100(26):15889–15894.

12 Sutherland, S. **2003** Ebola glycoprotein: the key to successful gene therapy? *Drug Discovery Today*; 8(14):609–610.

13 Simmons, G., Reeves, J.D, Grogan. C. C., et al. **2003** DC-SIGN and DC-SIGNR bind ebola glycoproteins and enhance infection of macrophages and endothelial cells. *Virology*; 305(1):115–123.

14 Schnittler, H. J. and Feldmann, H.**2003** Viral hemorrhagic fever – a vascular disease? *Thrombosis and Haemostasis*; 89(6):967–972.

15 Marz, A., Gramberg, .T, Simmon, G., et al. **2004** DC-SIGN and DC-SIGNR interact with the glycoprotein of Marburg virus and the S protein of severe acute respiratory syndrome coronavirus. *Journal of Virology*; 78(21):12090–12095.

16 Sullivan, N. J., Peterson, M., Yang, Z. Y., et al. **2005** Ebola virus glycoprotein toxicity is mediated by a dynamin-dependent protein-trafficking pathway. *Journal of Virology*; 79(1):547–553.

17 Volchkov, V.E., Feldmann, H., Volchkova, V.A, Klenk, H. D. **1998** Processing of the Ebola virus glycoprotein by the proprotein convertase furin. *Proceedings of the National Academy of Sciences of the United States of America*; 95(10):5762–5767.

18 Sanchez, A., Yang, Z. Y., Xu, L., et al. **1998** Biochemical analysis of the secreted and virion glycoproteins of Ebola virus. *Journal of Virology* ; 72(8):6442–6447.

19 Khan, A. S., Sanchez, A. and Pfleiger, A. **1998** Filoviral haemorrhagic fevers. *British Medical Bulletin*; 54:675–692.

20 Marburg Virus Disease. **1971** Berlin: Springer.

21 Smith, D. H., Johnson, B. K., Isaacson, M., et al. **1982** Marburg-virus disease in Kenya. *Lancet*; 1**8276**:816–820.

22 Anonymous. **2005** CDC Assists in Public Health Response to Marburg Hemorrhagic Fever Outbreak in Angola. In: Center for Disease Control.

23 Organization WH. **1978** Ebola Haemorrhagic fever in Zaire, 1976. Report of an international commission. *Bulletin of World Health Organization*; 56:271–293.

24 Organization WH. **1978** Ebola haemorrhagic fever in the Sudan, 1976. Report of a World Health Organization International Study Team. *Bulletin World Health Organization*; 56:247–270.

25 Organization WH. **1995** Ebola haemorrhagic fever – confirmed case in Cote d'Ivoire and suspect case in Leberia. *Weekly Epidemiological Record*; 70:359.

26 Jahrling, P. B., Geisbert, T. W., Dalgard, D. W., et al. **1990** Preliminary report: isolation of Ebola virus from monkeys imported to USA. *Lancet*; 335**8688**:502–505.

27 Anonymous. **1990** Update: Filovirus infection among persons with occupational exposure to nonhuman primates. MMWR Morbidity and Mortality Weekly Report; 39:266–273.

28 Swanepoel, R., Leman, P. A., Burt, F. J., et al. **1996** Experimental inoculation of plants and animals with Ebola virus. *Emerging Infectious Diseases*; 2(4):321–325.

29 Gear, J. S., Cassel, G.A, Gear, A. J., et al. **1975** Outbreak of Marburg virus disease in Johannesburg. *British Medical Journal*; 4**5995**:489–493.

30 Lacy, M. D. and Smego, R. A. **1996** Viral hemorrhagic fevers. *Advances in Pediatric Infectious Diseases*; 12:21–53.

31 Bwaka. M., Bonnet, M., Calain, P., et al. **1999** Ebola hemorrhagic fever in Kikwit, Democratic Republic of the Congo: Clinical observations in 103 patients. *Journal of Infectious Diseases*; 179(Suppl. 1):S1–S7.

32 Salvaggio, M. R. and Baddley, J. W. **2004** Other viral bioweapons: Ebola and Marburg hemorrhagic fever. *Dermatologic Clinics*; 22(3):291–302.

33 Sureau, P. H. **1989** Firsthand clinical observations of hemorrhagic manifestations in Ebola hemorrhagic fever in Zaire. *Reviews of Infectious Diseases* 11 Suppl 4:S790–S793.

34 Mahanty, S., Bray, M. **2004** Pathogenesis of filoviral haemorrhagic fevers. *The Lancet Infectious Diseases* 4(8):487–498.

35 Gill, V. Burke, A. **1995** Ebola hemorrhagic fever. *Infectious Disease Clinics of North America* 19:37–41.

36 Peters, C. and LeDuc, J. W. **1999** Ebola: The virus and the disease. *Journal of Infectious Diseases* 179(Suppl. 1):S1–S288.

37 Tomori, O., Bertolli, J., Rollin, P. E., et al. **1999** Serologic survey among hospital and health center workers during the Ebola hemorrhagic fever outbreak in Kikwit, Democratic Republic of the Congo, 1995. *Journal of Infectious Diseases* 179 Suppl 1:S98–S101.

38 Busico, K. M., Marshall, K. L., Ksiazek, T. G., et al. **1999** Prevalence of IgG antibodies to Ebola virus in individuals during an Ebola outbreak, Democratic Republic of the Congo, 1995. *Journal of Infectious Diseases* 179 Suppl 1:S102–S107.

39 Leroy, E. M., Baize, S., Volchkov, V.E., et al. **2000** Human asymptomatic Ebola infection and strong inflammatory response.[see comment]. *Lancet* 355**9222**:2210–2215.

40 Bazhutin, N., Belanov, E. F., Spiridonov, V., et al. **1992** The influence of the methods of experimental infection with Marburg virus on the course of illness in green monkeys. *Vopr Virusol* 37:153–156.

41 Belanov, Y., Muntyanov, V., Kryuk, V., et al. **1996** Retention of Marburg virus infecting capability on contaminated surfaces and in aersol particles. *Vopr Virusol* 41:32–34.

42 Peters, C. **2000** Are hemorrhagic fever viruses practical agents for biological terrorism? in *Scheld, Craig and Hughes Emerging Infections*. Washington, DC: ASM Press; 201–209.

43 Darling, R., Catlett, C., Huebner, K., et al. **2002** Threats in bioterrorism I: CDC category A agents. *Emergency Medicine Clinics of North America* 20:273–309.

44 Stroher, U., West, E., Bugany, H., et al. **2001** Infection and activation of monocytes by Marburg and Ebola viruses. *Journal of Virology* 75(22):11025–11033.

45 Geisbert, T. W., Young, H. A., Jahrling, P. B., et al. **2003** Pathogenesis of Ebola hemorrhagic fever in primate models: evidence that hemorrhage is not a direct effect of virus-induced cytolysis of endothelial cells. *American Journal of Pathology* 163(6):2371–2382.

46 Geisbert, T. W., Hensley, L. E., Larsen, T., et al. **2003** Pathogenesis of Ebola hemorrhagic fever in cynomolgus macaques: evidence that dendritic cells are early and sustained targets of infection. *American Journal of Pathology* 163(6):2347–2370.

47 Ksiazek, T. G., Rollin, P. E., Williams, A. J., et al. **1999** Clinical virology of Ebola hemorrhagic fever (EHF): virus, virus antigen, and IgG and IgM antibody findings among EHF patients in Kikwit, Democratic Republic of the Congo, 1995. *Journal of Infectious Diseases* 179 Suppl 1:S177–S187.

48 Towner, J. S., Rollin, P. E., Bausch, D. G., et al. **2004** Rapid diagnosis of Ebola hemorrhagic fever by reverse transcription-PCR in an outbreak setting and assessment of patient viral load as a predictor of outcome. *Journal of Virology* 78(8):4330–4341.

49 Sullivan, N., Yang, Z. Y., Nabel, G. J. **2003** Ebola virus pathogenesis: implications for vaccines and therapies. *Journal of Virology* 77(18):9733–9737.

50 Geisbert, T. W., Hensley, L. E., Gibb, T. R., et al. **2000** Apoptosis induced in vitro and in vivo during infection by Ebola and Marburg viruses. *Laboratory Investigation* 80(2):171–186.

51 Baize, S., Leroy, E. M., Georges-Courbot, M. C., et al. **1999** Defective humoral responses and extensive intravascular apoptosis are associated with fatal outcome in Ebola virus-infected patients.[see comment]. *Nature Medicine* 5(4):423–426.

52 Baize, S., Leroy, E. M., Mavoungou, E., et al. **2001** Apoptosis in fatal Ebola infection: Does the virus the bell for the immune system? *Apoptosis* 5:5–7.

53 Saccani, S., Pantano, S., Natoli, G. **2002** p38-Dependent marking of inflammatory genes for increased NF-kappa B recruitment.[see comment]. *Nature Immunology* 3(1):69–75.

54 Mahanty, S., Hutchinson, K., Agarwal, S., et al. **2003** Cutting edge: impairment of dendritic cells and adaptive immunity by Ebola and Lassa viruses. *Journal of Immunology* 170(6):2797–2801.

55 Basler, C. F., Wang, X., Muhlberger, E., et al. **2000** The Ebola virus VP35 protein functions as a type I IFN antagonist. *Proceedings of the National Academy of Sciences of the United States of America* 97(22):12289–12294.

56 Basler, C. F., Mikulasova, A., Martinez-Sobrido, L., et al. **2003** The Ebola virus VP35 protein inhibits activation of interferon regulatory factor 3. *Journal of Virology* 77(14):7945–7956.

57 Bosio, C. M., Aman, M. J., Grogan, C., et al. **2003** Ebola and Marburg viruses replicate in monocyte-derived dendritic cells without inducing the production of cytokines and full maturation.[see comment]. *Journal of Infectious Diseases* 188(11):1630–1638.

58 Geisbert, T. W., Hensley, L. E., Jahrling, P. B., et al. **2003** Treatment of Ebola virus infection with a recombinant inhibitor of factor VIIa/tissue factor: a study in rhesus monkeys. *Lancet* 362**9400**:1953–1958.

59 Geisbert, T. W., Young, H. A., Jahrling, P. B., et al. **2003** Mechanisms underlying coagulation abnormalities in ebola hemorrhagic fever: overexpression of tissue factor in primate monocytes/macrophages is a key event.[see comment]. *Journal Diseases* 188(11):1618–1629.

60 Ruf, W. **2004** Emerging roles of tissue factor in viral hemorrhagic fever. *Trends in Immunology* 25(9):461–464.

61 Wahl-Jensen, V., Kurz, S. K., Hazelton, P. R., et al. **2005** Role of Ebola virus secreted glycoproteins and virus-like particles in activation of human macrophages. *Journal of Virology* 79(4):2413–2419.

62 Villinger, F., Rollin, P. E., Brar, S. S., et al. **1999** Markedly elevated levels of interferon (IFN)-gamma, IFN-alpha, interleukin (IL)-2, IL-10, and tumor necrosis factor-alpha associated with fatal Ebola virus infection. *Journal of Infectious Diseases* 179 Suppl 1:S188–S191.

63 Baize, S., Leroy, E. M., Georges, A. J., et al. **2002** Inflammatory responses in Ebola virus-infected patients. *Clinical and Experimental Immunology* 128(1):163–168.

64 Leroy, E. M., Baize, S., Debre, P., et al. **2001** Early immune responses accompanying human asymptomatic Ebola infections. *Clinical and Experimental Immunology* 124(3):453–460.

65 Warfield, K. L., Olinger, G., Deal, E. M., et al. **2005** Induction of humoral and CD8+ T cell responses are required for protection against lethal Ebola virus infection. *Journal of Immunology* 175(2):1184–1191.

66 Ksiazek, T. G., West, C. P., Rollin, P. E., et al. **1999** ELISA for the detection of antibodies to Ebola viruses. *Journal of Infectious Diseases* 179 Suppl 1:S192–S198.

67 Clegg, J. C., Wilson, S. M., Oram, J. D. **1991** Nucleotide sequence of the S RNA of Lassa virus (Nigerian strain) and comparative analysis of arenavirus gene products. *Virus Research* 18(2/3):151–164.

68 Griffiths, C. M., Wilson, S. M., Clegg, J. C. **1992** Sequence of the nucleocapsid protein gene of Machupo virus: close relationship with another South American pathogenic arenavirus, Junin. *Archives of Virology* 124(3/4):371–377.

69 Bowen, M. D., Peters, C. J., Nichol, S. T. **1996** The phylogeny of New World (Tacaribe complex) arenaviruses. *Virology* 219(1):285–290.

70 Gonzalez, J. P., Bowen, M. D., Nichol, S. T., et al. **1996** Genetic characterization and phylogeny of Sabia virus, an emergent pathogen in Brazil. *Virology* 221(2):318–324.

71 Bowen, M. D., Peters, C. J., Nichol, S. T. **1997** Phylogenetic analysis of the Arenaviridae: patterns of virus evolution and evidence for cospeciation between arenaviruses and their rodent hosts. *Molecular Phylogenetics and Evolution* 8(3):301–316.

72 Albarino, C. G., Posik, D. M., Ghiringhelli, P. D., et al. **1998** Arenavirus phylogeny: a new insight. *Virus Genes* 16(1):39–46.

73 Fulhorst, C. F., Bowen, M. D., Salas, R. A., et al. **1999** Natural rodent host associations of Guanarito and pirital viruses (Family Arenaviridae) in central Venezuela. *American Journal of Tropical Medicine and Hygiene* 61(2):325–330.

74 Bowen, M. D., Rollin, P. E., Ksiazek, T. G., et al. **2000** Genetic diversity among Lassa virus strains. *Journal of Virology* 74(15):6992–7004.

75 Weaver, S. C., Salas, R. A., de Manzione, N., et al. **2000** Guanarito virus (Arenaviridae) isolates from endemic and outlying localities in Venezuela: sequence comparisons among and within strains isolated from Venezuelan hemorrhagic fever patients and rodents. *Virology* 266(1):189–195.

76 Hugot, J. P., Gonzalez, J. P., Denys, C. **2001** Evolution of the Old World Arenaviridae and their rodent hosts: generalized host-transfer or association by descent? *Infection, Genetics and Evolution* 1(1):13–20.

77 Spiropoulou, C. F., Kunz, S., Rollin, P. E., et al. **2002** New World arenavirus clade C, but not clade A and B viruses, utilizes alpha-dystroglycan as its major receptor. *Journal of Virology* 76(10):5140–5146.

78 Gunthe, S., Lenz, O. **2004** Lassa virus. *Critical Reviews in Clinical Laboratory Sciences* 41(4):339–390.

79 Cajimat, M. N., Fulhorst, C. F. **2004** Phylogeny of the Venezuelan arenaviruses. *Virus Research* 102(2):199–206.

80 Buchmeier, M., Bowen, M., Peters, C. **2001** Arenaviridae: The viruses and their replication in *Knipe and Howleys Fields Virology*. 4th ed. Philadelphia: Lippincott Williams and Wilkens 1635–1668.

81 Monath, T. P. **1975** Lassa fever: Review of epidemiology of and epizootology. *Bulletin World Health Organization* 52:577.

82 McCormick, J. B., Webb, P. A., Krebs, J. W., et al. **1987** A prospective study of the epidemiology and ecology of Lassa fever. *Journal of Infectious Diseases* 155(3):437–444.

83 Peters, C. **2005** Lymphocytic choriomeningits virus, Lassa virus, and the South American Hemorrhagic Fevers in *Mandell, Bennett and Dolins Principles and Practices of Infectious Diseases*. 6th ed. Philadelphia: Elsevier Churchill Livingstone 2090–2098.

84 McCormick, J. B., King, I. J., Webb, P. A., et al. **1987** A case-control study of the clinical diagnosis and course of Lassa fever. *Journal of Infectious Diseases* 155(3):445–455.

85 Monath, T. P. **1973** Lassa fever. *Tropical Doctor* 3(4):155–161.

86 Monath, T. P., Maher, M., Casals, J., et al. **1974** Lassa fever in the Eastern Province of Sierra Leone, 1970–1972. II. Clinical observations and virological studies on selected hospital cases. *American Journal of Tropical Medicine and Hygiene* 23(6):1140–1149.

87 Monath, T. P., Casals, J. **1975** Diagnosis of Lassa fever and the isolation and management of patients. *Bulletin of the World Health Organization* 52(4/6):707–715.

88 Johnson, K. M., McCormick, J. B., Webb, P. A., et al. Clinical virology of Lassa fever in hospitalized patients. *Journal of Infectious Diseases* 155(3):456–464.

89 Cummins, D., McCormick, J. B., Bennett, D., et al. **1990** Acute sensorineural deafness in Lassa fever.[see comment]. *JAMA* 264(16):2093–2096.

90 Cummins, D., Bennett, D., Fisher-Hoch, S. P., et al. **1992** Lassa fever encephalopathy: clinical and laboratory findings. *Journal of Tropical Medicine and Hygiene* 95(3):197–201.

91 Weissenbacher, M. C., Sabattini, M. S., Avila, M. M., et al. **1983** Junin virus activity in two rural populations of the Argentine hemorrhagic fever (AHF) endemic area. *Journal of Medical Virology* 12(4):273–280.

92 Kilgore, P. E., Ksiazek, T. G., Rollin, P. E., et al. **1997** Treatment of Bolivian hemorrhagic fever with intravenous ribavirin. *Clinical Infectious Diseases* 24(4):718–722.

93 Tesh, R. B. **2002** Viral hemorrhagic fevers of South America. *Biomedica* 22(3):287–295.

94 Charrel, R. N., de Lamballerie, X. **2003** Arenaviruses other than Lassa virus. *Antiviral Research* 57(1/2):89–100.

95 Price, M. E., Fisher-Hoch, S. P., Craven, R. B., et al. **1988** A prospective study of maternal and fetal outcome in acute Lassa fever infection during pregnancy. *BMJ* 297**6648**:584–587.

96 Maiztegui, J. I., Fernandez, N. J., de Damilano, A. J. **1979** Efficacy of immune plasma in treatment of Argentine haemorrhagic fever and association between treatment and a late neurological syndrome. *Lancet* 28**154**:1216–1217.

97 Enria, D. A., Briggiler, A. M., Fernandez, N. J., et al. **1984** Importance of dose of neutralising antibodies in treatment of Argentine haemorrhagic fever with immune plasma. *Lancet* 28**397**:255–256.

98 Harrison, L. H., Halsey, N. A., McKee, K. T., Jr., et al. **1999** Clinical case definitions for Argentine hemorrhagic fever. *Clinical Infectious Diseases* 28(5):1091–1094.

99 Enria, D. A., Maiztegui, J. I. **1994** Antiviral treatment of Argentine hemorrhagic fever. *Antiviral Research* 23(1):23–31.

100 McCormick, J. B., King, I. J., Webb, P. A., et al. **1986** Lassa fever. Effective therapy with ribavirin. *New England Journal of Medicine* 314(1):20–26.

101 Petkevich, A. S., Sabynin, V. M., Lukashevich, I. S., et al. **1981** Vliianie ribovirina (virazola) na reproduktsiiu nekotorykh arenavirusov v kul'ture kletok. *Voprosy Virusologii* (2):244–245.

102 Rodriguez, M., McCormick, J. B., Weissenbache,r M. C. **1986** Antiviral effect of ribavirin on Junin virus replication in vitro. *Revista Argentina de Microbiologia* 18(2):69–74.

103 Weissenbacher, M. C., Calello, M. A., Merani, M. S., et al. **1986** Therapeutic effect of the antiviral agent ribavirin in Junin virus infection of primates. *Journal of Medical Virology* 20(3):261–267.

104 Weissenbacher, M. C., Avila, M. M., Calello, M. A., et al. **1986** Effect of ribavirin and immune serum on Junin virus-infected primates. *Medical Microbiology and Immunology* 175(2/3):183–186.

105 Kenyon, R. H., Canonico, P. G., Green, D. E., et al. **1986** Effect of ribavirin and tributylribavirin on argentine hemorrhagic fever (Junin virus) in guinea pigs. *Antimicrobial Agents and Chemotherapy* 129(3):521–523.

106 McKee, K. T., Jr., Huggins, J. W., Trahan, C. J., et al. **1988** Ribavirin prophylaxis and therapy for experimental argentine hemorrhagic fever. *Antimicrobial Agents and Chemotherapy* 32(9):1304–1309.

107 Remesar, M. C., Blejer, J. L., Weissenbacher, M. C., et al. **1988** Ribavirin effect on experimental Junin virus-induced encephalitis. *Journal of Medical Virology* 26(1):79–84.

108 Moss, J. T., Wilson, J. P. **1992** Treatment of viral hemorrhagic fevers with ribavirin. *Annals of Pharmacotherapy* 26(9):1156–1157.

109 Maiztegui, J. I. McKee, K. T., Jr., Barrera-Oro, J. G., et al. **1998** Protective efficacy of a live attenuated vaccine against Argentine hemorrhagic fever. AHF Study Group. *Journal of Infectious Diseases* 177(2):277–283.

110 Peters, C. J., Kuehne, R. W., Mercado, R. R., et al. **1974** Hemorrhagic fever in Cochabamba, Bolivia, 1971. *American Journal of Epidemiology* 99(6):425–433.

111 Ambrosio, A. M., Enria, D. A., Maiztegui, J. I. **1986** Junin virus isolation from lympho-mononuclear cells of patients with Argentine hemorrhagic fever. *Intervirology* 25(2):97–102.

112 Park, J. Y., Peters, C. J., Rollin, P. E., et al. **1997** Development of a reverse transcription-polymerase chain reaction assay for diagnosis of lymphocytic choriomeningitis virus infection and its use in a prospective surveillance study. *Journal of Medical Virology* 51(2):107–114.

113 Schmalijohn, C., Hooper, J. W. **2001** Bunyaviridae: The Viruses and Their Replication. in *Knipe and Howleys Fields Virology*. Philadelphia: Lippincott Williams and Wilkins 1581–1633.

114 Capstick, P. B., Gosden, D. **1962** Neutralizing antibody response of sheep to pantropic and neutrotropic rift valley fever virus. *Nature* 195:583–584.

115 Lewis, J. C., Botros, B. A., Meegan, J. M. **1978** Studies on Rift Valley fever in the domestic animals in Egypt. *Journal of the Egyptian Public Health Association* 53(3/4):271–272.

116 Imam, I. Z., Karamany, R. E., Darwish, M. A. **1978** Epidemic of Rift Valley fever (RVF) in Egypt: isolation of RVF virus from animals. *Journal of the Egyptian Public Health Association* 53(3/4):265–269.

117 Ali, A. M., Kamel, S. **1978** Epidemiology of RVF in domestic animals in Egypt. *Journal of the Egyptian Public Health Association* 53(3/4):255–263.

118 Meegan, J. M., Moussa, M. I. **1978** Viral studies of Rift Valley fever in the Arab Republic of Egypt. *Journal of the Egyptian Public Health Association* 53(3/4):243–244.

119 Imam, I. Z., Darwish, M. A., Karamany, R. E. **1978** Epidemic of Rift Valley fever (RVF) in Egypt: virological diagnosis of RVF in man. *Journal of the Egyptian Public Health Association* 53(3/4):205–208.

120 Wahab, K. S., el-Baz, L. M., el-Tayeb, E. M., et al. **1978** Virus isolation and identification from cases of Rift Valley fever virus infection in Egypt. *Journal of the Egyptian Public Health Association* 53(3/4):201–203.

121 Hjelle, B., Jenison, S., Torrez-Martinez, N., et al. **1994** A novel hantavirus associated with an outbreak of fatal respiratory disease in the southwestern United States: evolutionary relationships to known hantaviruses. *Journal of Virology* 68(2):592–596.

122 Hjelle, B., Jenison, S., Mertz, G., et al. **1994** Emergence of hantaviral disease in the southwestern United States. *Western Journal of Medicine* 161(5):467–473.

123 Morrison, Y. Y., Rathbun, R. C. **1995** Hantavirus pulmonary syndrome: the Four Corners disease. *Annals of Pharmacotherapy* 29(1):57–65.

124 Duchin, J. S., Koster, F. T., Peters, C. J., et al. **1994** Hantavirus pulmonary syndrome: a clinical description of 17 patients with a newly recognized disease. The Hantavirus Study Group.[see comment]. *New England Journal of Medicine* 330(14):949–955.

125 Moolenaar, R. L., Dalton, C., Lipman, H. B., et al. **1995** Clinical features that differentiate hantavirus pulmonary syndrome from three other acute respiratory illnesses. *Clinical Infectious Diseases* 21(3):643–649.

126 Peters, C. J., Simpson, G. L., Levy, H. **1999** Spectrum of hantavirus infection: hemorrhagic fever with renal syndrome and hantavirus pulmonary syndrome. *Annual Review of Medicine* 50:531–545.

127 Moolenaar, R. L., Breiman, R. F., Peters, C. J. **1997** Hantavirus pulmonary syndrome. *Seminars in Respiratory Infections* 12(1):31–39.

128 Peters, C. J., Khan, A. S. **2002** Hantavirus pulmonary syndrome: the new American hemorrhagic fever. *Clinical Infectious Diseases* 34(9):1224–1231.

129 Chan, Y. C., Wong, T. W., Yap, E. H. **1987** Haemorrhagic fever with renal syndrome: clinical, virological and epidemiological perspectives. *Annals of the Academy of Medicine, Singapore* 16(4):696–701.

130 Yao, Z. O., Yang, W. S., Zhang, W. B., et al. **1989** The distribution and duration of hantaan virus in the body fluids of patients with hemorrhagic fever with renal syndrome. *Journal of Infectious Diseases* 160(2):218–224.

131 Niklasson, B. S. **1992** Haemorrhagic fever with renal syndrome, virological and epidemiological aspects. *Pediatric Nephrology* 6(2):201–204.

132 Ko, K. W. **1992** Haemorrhagic fever with renal syndrome: clinical aspects. *Pediatric Nephrology* 6(2):197–200.

133 Escutenaire, S., Pastoret, P. P. **2000** Hantavirus infections. *Revue Scientifique et Technique* 19(1):64–78.

134 Peters, C. J. **2005** California Encephalitis, Hantavirus Pulmonary Syndrome and Bunyavirid Hemorrhagic Fevers. in *Mandell, Bennett, and Dolins Principles and Practice of Infectious Diseases*. Philadelphia: Elsevier Churchill Livingstone 2086–2090.

135 Huggins, J. W., Hsiang, C. M., Cosgriff, T. M., et al. **1991** Prospective, double-blind, concurrent, placebo-controlled clinical trial of intravenous ribavirin therapy of hemorrhagic fever with renal syndrome. *Journal of Infectious Diseases* 164(6):1119–1127.

136 Yassin, W. **1978** Clinico-pathological picture in five human cases died with Rift Valley fever. *Journal of the Egyptian Public Health Association* 53(3/4):191–193.

137 Laughlin, L. W., Girgis, N. I., Meegan, J. M., et al. **1978** Clinical studies on Rift Valley fever. Part 2: Ophthalmologic and central nervous system complications. *Journal of the Egyptian Public Health Association* 53(3/4):183–184.

138 Strausbaugh, L. J., Laughlin, L. W., Meegan, JM., et al. **1978** Clinical studies on Rift Valley fever, Part I: Acute febrile and hemorrhagic-like diseases. *Journal of the Egyptian Public Health Association* 53(3/4):181–182.

139 Boctor, W. M. **1978** The clinical picture of Rift Valley fever in Egypt. *Journal of the Egyptian Public Health Association* 53(3/4):177–180.

140 Meegan, J. M., Moussa, M. I., el-Mour, A. F., et al. **1978** Ecological and epidemiological studies of Rift Valley fever in Egypt. *Journal of the Egyptian Public Health Association* 53(3/4):173–175.

141 Abdel-Wahab, K. S., El-Baz, L. M., El-Tayeb, E. M., et al. **1978** Rift Valley Fever virus infections in Egypt: Pathological and virological findings in man. *Transactions of the Royal Society of Tropical Medicine and Hygiene* 72(4):392–396.

142 Johnson, B. K., Chanas, A. C., el-Tayeb, E., et al. **1978** Rift Valley fever in Egypt, 1978. *Lancet* 2(8092 Pt 1):745.

143 Meegan, J. M. **1979** The Rift Valley fever epizootic in Egypt 1977–78. 1. Description of the epizzotic and virological studies. *Transactions of the Royal Society of Tropical Medicine and Hygiene* 73(6):618–623.

144 Al-Hazmi, M., Ayoola, E. A., Abdurahman, M., et al. **2003** Epidemic Rift Valley fever in Saudi Arabia: a clinical study of severe illness in humans. *Clinical Infectious Diseases* 36(3):245–252.

145 Siam, A.L, Gharbawi, K. F., Meegan, J. M. **1978** Ocular complications of Rift Valley fever. *Journal of the Egyptian Public Health Association* 53(3/4):185–186.

146 Siam, A. L., Meegan, J. M. **1980** Ocular disease resulting from infection with Rift Valley fever virus. *Transactions of the Royal Society of Tropical Medicine and Hygiene* 74(4):539–541.

147 Kark, J. D., Aynor, Y., Peters, CJ. **1982** A Rift Valley fever vaccine trial. I. Side effects and serologic response over a six-month follow-up. *American Journal of Epidemiology* 116(5):808–820.

148 Kark, J. D., Aynor, Y., Peters, C. J. **1985** A Rift Valley fever vaccine trial: 2. Serological response to booster doses with a comparison of intradermal versus subcutaneous injection. *Vaccine* 3(2):117–122.

149 Meadors, G. F., 3rd, Gibbs, P. H., Peters, C. J. **1986** Evaluation of a new Rift Valley fever vaccine: safety and immunogenicity trials. *Vaccine* 4(3):179–184.

150 Mayers, D. L. **1999** Exotic virus infections of military significance. *Dermatologic Clinics* 17:29–40.

151 Swanepoel, R., Shepherd, A. J., Leman, P. A., et al. **1987** Epidemiologic and clinical features of Crimean–Congo hemorrhagic fever in southern Africa. *American Journal of Tropical Medicine and Hygiene* 36(1):120–132.

152 Mayers, D. L. **1999** Exotic virus infections of military significance. Hemorrhagic fever viruses and pox virus infections. *Dermatologic Clinics* 17(1):29–40.

153 van de Wal, B. W., Joubert, J. R., van Eeden, P. J., et al. **1985** A nosocomial outbreak of Crimean–Congo haemorrhagic fever at Tygerberg Hospital. Part IV. Preventive and prophylactic measures. *South African Medical Journal. Suid-Afrikaanse Tydskrif Vir Geneeskunde* 68(10):729–732.

154 van Eeden, P. J., van Eeden, S. F., Joubert, J. R., et al. **1985** A nosocomial outbreak of Crimean–Congo haemorrhagic fever at Tygerberg Hospital. Part II. Management of patients. *South African Medical Journal. Suid-Afrikaanse Tydskrif Vir Geneeskunde* 68(10):718–721.

155 Tignor, G. H., Hanham, C. A. **1993** Ribavirin efficacy in an in vivo model of Crimean–Congo hemorrhagic fever virus (CCHF) infection. *Antiviral Research* 22(4):309–325.

156 Fisher-Hoch, S. P., Khan, J. A., Rehman, S., et al. **1995** Crimean–Congo haemorrhagic fever treated with oral ribavirin. *Lancet* 346**8973**:472–475.

157 Bangash, S. A., Khan, E. A. **2003** Treatment and prophylaxis with ribavirin for Crimean–Congo Hemorrhagic Fever – is it effective? *JPMA – Journal of the Pakistan Medical Association* 53(1):39–41.

158 Whitehouse, C. A. **2004** Crimean–Congo hemorrhagic fever. *Antiviral Research* 64(3):145–160.

159 Ergonul, O., Celikbas, A., Dokuzoguz, B., et al. **2004** Characteristics of patients with Crimean–Congo hemorrhagic fever in a recent outbreak in Turkey and impact of oral ribavirin therapy. *Clinical Infectious Diseases* 39(2):284–287.

160 Bakir, M., Ugurlu, M., Dokuzoguz, B., et al. **2005** Crimean–Congo haemorrhagic fever outbreak in Middle Anatolia: a multicentre study of clinical features and outcome measures. *Journal of Medical Microbiology* 54(Pt 4):385–389.

161 Lee, H. W., Baek, L. J., Johnson, K. M. **1982** Isolation of Hantaan virus, the etiologic agent of Korean hemorrhagic fever, from wild urban rats. *Journal of Infectious Diseases* 146(5):638–644.

162 Anonymous. **1982** Preliminary evidence that Hantaan or a closely related virus is enzootic in domestic rodents. *New England Journal of Medicine* 307(10):623–625.

163 LeDuc, J. W. **1987** Epidemiology of Hantaan and related viruses. *Laboratory Animal Science* 37(4):413–418.

164 Wong, T. W., Chan, Y. C., Joo, Y. G., et al. **1989** Hantavirus infections in humans and commensal rodents in Singapore. *Transactions of the Royal Society of Tropical Medicine and Hygiene* 83(2):248–251.

165 Enria, D. A., Pinheiro, F. **2000** Rodent-borne emerging viral zoonosis. Hemorrhagic fevers and hantavirus infections in South America. *Infectious Disease Clinics of North America* 14(1):167–184.

166 Fritz, C. L., Fulhorst, C. F., Enge, B., et al. **2002** Exposure to rodents and rodent-borne viruses among persons with elevated occupational risk. *Journal of Occupational and Environmental Medicine* 44(10):962–967.

167 Pini, N., Levis, S., Calderon, G., et al. **2003** Hantavirus infection in humans and rodents, northwestern Argentina. *Emerging Infectious Diseases* 9(9):1070–1076.

168 Papa, A., Bino, S., Llagami, A., et al. **2002** Crimean–Congo hemorrhagic fever in Albania, 2001. *European Journal of Clinical Microbiology and Infectious Diseases* 21(8):603–606.

169 Brown, J. L., Dominik, J. W., Morrissey, R. L. **1981** Respiratory infectivity of a recently isolated Egyptian strain of Rift Valley fever virus. *Infection and Immunity* 33(3):848–853.

170 Lee, H. W., Johnson, K. M. **1982** Laboratory-acquired infections with Hantaan virus, the etiologic agent of Korean hemorrhagic fever. *Journal of Infectious Diseases* 146(5):645–651.

171 Nuzum, E. O., Rossi, C. A., Stephenson, E. H., et al. **1988** Aerosol transmission of Hantaan and related viruses to laboratory rats. *American Journal of Tropical Medicine and Hygiene* 38(3):636–640.

172 Anonymous. **1975** The new program of the World Health Organization in medical virology. *Intervirology* 6(3):133–149.

173 Harxhi, A., Pilaca, A., Delia, Z., et al. **2005** Crimean–Congo hemorrhagic fever: a case of nosocomial transmission. *Infection* 33(4):295–296.

174 Bhatt, P. N., Work, T. H., Varma, M. G., et al. **1966** Kyasanur forest diseases. IV. Isolation of Kyasanur forest disease virus from infected humans and monkeys of Shimogadistrict, Mysore state. *Indian Journal of Medical Sciences* 20(5):316–320.

175 Boshell, J., Rajagopalan, P. K., Patil, A. P., et al.1968 Isolation of Kyasanur Forest disease virus from ixodid ticks: 1961–1964. *Indian Journal of Medical Research* 56(4):541–568.

176 Boshell, J., Rajagopalan, P. K., Goverdhan, M. K., et al. **1968** The isolation of Kyasanur Forest disease virus from small mammals of the Sagar–Sorab forests, Nysore State, India: 1961–1964. *Indian Journal of Medical Research* 56(4):569–572.

177 Pavri, K. **1989** Clinical, clinicopathologic, and hematologic features of Kyasanur Forest disease. *Reviews of Infectious Diseases* 11 Suppl 4:S854–S859.

178 Achar, T. R., Patil, A. P., Jayadevaiah, M. S. **1981** Persistence of humoral immunity in Kyasanur forest disease. *Indian Journal of Medical Research* 73:1–3.

179 Dandawate, C. N., Mansharamani, H. J., Jhala, H. I. **1965** Experimental vaccine against Kyasanur Forest Disease (KFD) virus from embryonated eggs. I. Adaptation of the virus to developing chick embryo and preparation of formolised vaccines. *Indian Journal of Pathology and Bacteriology* 8(4):241–260.

180 Dandawate, C. N., Mansharamani, H. J., Jhala, H. I. **1965**Experimental vaccine against Kyasanur Forest Disease (KFD) virus from embryonated eggs. II. Comparative immunogenicity of vaccines inactivated with formalin and beta-propio-lactone (BPL). *Indian Journal of Pathology and Bacteriology* 8(4):261–270.

181 Kharitonova, N., Yu, A. **1985** *Omsk hemorrhagic fever. Ecology of the Agent and Epizootiology.* New Dehli: Amerind Publishing Company.

182 Gritsun, T. S., Nuttall, P. A., Gould, E. A. **2003** Tick-borne flaviviruses. *Advances in Virus Research* 61:317–371.

183 World Health Organization **1970** Health Aspects for Chemical and Biological Weapons: Report of a Group of Consultants. in Geneva, Switzerland: World Health Organization 1–132.

184 Letson, G. W., Bailey, R. E., Pearson, J., et al. **1993** Eastern equine encephalitis (EEE): a description of the 1989 outbreak, recent epidemiologic trends, and the association of rainfall with EEE occurrence. *American Journal of Tropical Medicine and Hygiene* 49(6):677–685.

185 Bigler, W. J., Lassing, E. B., Buff, E. E., et al. **1976** Endemic eastern equine encephalomyelitis in Florida: a twenty-year analysis, 1955–1974. *American Journal of Tropical Medicine and Hygiene* 25(6):884–890.

186 Przelomski, M. M., O'Rourke, E., Grady, G. F., et al. **1988** Eastern equine encephalitis in Massachusetts: a report of 16 cases, 1970–1984. *Neurology* 38(5):736–739.

187 Baker, A., II. **1958** Western equine encephalitis. Clinical features. *Neurology* 8:880–881.

188 Koprowski, H., Cox, H. **1947** Human laboratory infection with Venezuelan equine encephalomyelitis. *New England Journal of Medicine* 236:647.

189 Tsai, T., Monath, T. P. **1997** Alphaviruses. in *Richman, Whitley and Haydens Clinical virology.* New York: Churchill Livingstone; 1217–1255.

10

Policy Priorities: Smallpox, Stockpiles, and Surveillance

Ross D. Silverman

10.1
Introduction

The United States public health legal system is as diverse and decentralized as the United States public health system itself. In the context of bioterrorism, it is of critical importance that those empowered to act in response to a public health threat clearly understand the powers available to them and other personnel under the law, and the gaps in such laws, otherwise efforts to effectively utilize this assortment of public health protections and programs will probably fail. In this chapter and the chapter entitled "Legal Preparedness: The Modernization Of State, National And International Public Health Law" we examine the lessons learned from several federal policy initiatives, including the pre-event smallpox vaccination program begun in January 2003, and the creation of new national stockpile and surveillance programs to assist preparedness efforts; the legal foundations of bioterrorism-related public health authority; the process of ensuring legal preparedness of the public health system; and the reform of state statutes in an effort to improve state responsiveness to public health emergencies.

10.2
Smallpox Preparedness and Pre-event Vaccination

One of the most instructive examples of the legal, ethical, political, and logistical complexities underlying the national public health preparedness policymaking process can be found in the Administration's effort to develop and implement a national smallpox pre-event vaccination program.

The anthrax attacks of October 2001, coupled with the terrorist attacks of September 11, 2001, led many to raise concerns about the possibility of a large-scale biological attack on the United States, with smallpox virus being the most

Bioterrorism Preparedness. Edited by Nancy Khardori
Copyright © 2006 WILEY-VCH Verlag GmbH & Co. KGaA, Weinheim
ISBN: 3-527-31235-8

feared weapon. Although only two countries – the United States and Russia – were officially recognized as possessing stores of the smallpox virus, several other countries, including North Korea and Iraq, were believed to possess covert stockpiles of the deadly pathogen [1, 2]. In late 2001 the Bush Administration requested from Congress $509 million to be used to purchase a stockpile of 300 million smallpox vaccine doses over the course of the following year. This action started a wide-ranging public debate over smallpox and smallpox preparedness which asked the questions:

- What were the real risks of an attack?
- How vulnerable was the American public both to an attack and to the vaccine?
- What was the best way to defend against smallpox should an attack occur or be forecast?
- Who should receive smallpox vaccinations, and how and when [3–5]?

As the public and the public health sector debated these questions the Administration turned to the CDC's Advisory Committee on Immunization Practices (ACIP) and a Smallpox Working Group comprising ACIP members and other scientific experts for guidance on several issues – the relative risks of vaccination compared with the risks of attack, who (if anyone) should receive vaccination as part of a pre-event vaccination strategy, and the best approach with which to respond to the appearance of a smallpox case. The ACIP was faced with a difficult task – smallpox had not been seen in the United States for over three decades and much about the American population had changed (for example 21st century America had a significantly higher population of immunocompromised and unvaccinated people); available data was therefore limited in its usefulness. Furthermore, little was known about, or disclosed to, the ACIP about the real risks of an attack, although it was generally believed that the risk of attack, while serious, was remote. Consequently, it was not surprising that in June 2002, the ACIP recommended against pre-event vaccination of the general public and recommended a limited number of local, state, and federal first responders (10,000–20,000) be vaccinated as part of a first-response strategy which would rely upon the use of surveillance and containment by ring vaccination [6]. Over the course of the summer and autumn, public debate continued while federal officials refined their smallpox-response plans. Advisors to the Department of Health and Human Services, including D. A. Henderson, urged expansion of the number of pre-event first responders vaccinated. In September, as the Administration increased the heat of its rhetoric concerning the threat posed to the United States by Iraq, the Department of Health and Human Services publicly announced its desire to vaccinate between 250,000 and 500,000 first responders as part of a pre-event vaccination strategy, in stark contrast with the ACIP recommendation earlier that year; this resulted in additional questions and confusion about the true risks of a smallpox attack [7]. During the autumn of 2002 the CDC requested the establishment and advice of the Institute of Medicine's Committee on Smallpox Vaccination Program Implementation (IOM Committee); during the course of the following year this committee released six reports which reviewed the medical, political, legal, and practical challenges faced by policymakers and the US public health community during the implementation of the pre-event smallpox vaccination program [8–13]. The CDC also worked with state and local public health leaders to prepare smallpox-response plans, although details about the risks of attack and what would ultimately be recommended by the administration remained vague (in fact, it became known that administration officials did not share its evidence of smallpox attack risk with its advisory panel [1]). Furthermore, the remote chances of receiving additional funding to assist with local smallpox preparedness efforts raised additional questions about the feasibility of a broad initiative.

On December 13, 2002, President Bush announced his pre-event smallpox vaccination plan for military and civilian populations [14]. More than 500,000 members of the military would receive smallpox vaccination, beginning immediately. The three phase civilian pre-event vaccination plan would begin in January 2003. Phase I called for the vaccination of 500,000 volunteer public health and health care workers; in Phase II, 10 million additional health care, public health and emergency responders would be vaccinated; and at some point after the safe vaccination of the second wave of emergency responders the administration would weigh the benefits of making the vaccine available to the general public.

In many ways this announcement overwhelmed state public-health departments. While the President asserted "there is no reason to believe that smallpox presents an imminent threat," his proposal marked a dramatic expansion of the number to be vaccinated pre-event (from at most 500,000 to 10.5 million) and presented the departments with a hurried timetable under which they were expected to have their programs up and running. The estimated costs the already underfunded [15] local and state health departments were expected to bear in this initiative were substantial – between $600 million and $1 billion [16] – and most of the federal preparedness dollars which had been distributed to state health departments had already been put to use to address issues such as anthrax response, West Nile virus, and (in early 2003) SARS. It was inevitable that essential public health services would be sacrificed to respond to this new initiative [17], and that comprehensive preparedness efforts focused on achieving necessary improvements in the overall public health infrastructure would be subsumed to the now-immediate needs of a program geared to respond to a particular biological agent [11].

The IOM Committee later described the complications of the smallpox plan as:

> [A]n atypical vaccination campaign...a public health component
> of bioterrorism preparedness...[in which public health authorities
> are being asked to] implement a program with inherent serious
> risks and with publicly unknown and unstated benefits, and to do
> so rapidly, within a timeline that has not been explicitly outlined.

Other experts described the test posed to public health authorities by the new directive as follows:

> When the public health sector was informed, rather precipitously,
> that it needed to redirect its preparedness efforts to a specific
> smallpox initiative, they had "less than 3 weeks to develop their

> *plans and less than two months to prepare to begin vaccination."*
> *Preparations included developing and submitting "pre-attack"*
> *vaccination plans to the CDC that covered the size of each*
> *Smallpox Response Team; the location of each vaccination site; the*
> *number of health care facilities identified to participate in the*
> *Phase I program; information on vaccine logistics and security,*
> *training, and data management; and other key aspects of their*
> *plans. Public health was in its usual position of doing more than it*
> *was funded to do with very little notice, but in the unusual position*
> *of attempting to do it under the lens of the nation. Furthermore,*
> *the sudden and rather onerous amount of work and the plans on*
> *which their deliverables would be based was placed before the*
> *public health sector without benefit of smallpox program-specific*
> *funding* [18].

Many health care professional associations, prominent hospital systems, and public health leaders questioned both the need for the program and publicly announced that they would not participate or encourage others to participate in the program [19].

An official declaration by the Secretary of Health and Human Services in late January 2003 [20] marked the formal start of the civilian vaccination program. However, largely because of significant concerns about the physical and economic risks posed by vaccination (see discussion below), and little new evidence of a risk of smallpox attack arising during the buildup and preliminary stages of the war in Iraq, far fewer hospitals, public-health departments, and health care workers than desired were willing to participate, and by the middle of June 2003, the military and civilian vaccination programs "had virtually come to a halt" [21]. By January 31, 2005, 39,608 civilian health-care personnel had been vaccinated as part of the smallpox vaccination program [22], approximately 8% of the Phase I goal stated two years earlier.

Why did the pre-event smallpox vaccination program fail?

In addition to the myriad logistical, ethical and scientific questions raised in the discussion above, some of the most significant obstacles to the success of the smallpox program were legal in nature. Specifically, the commencement of the smallpox vaccination plan was fatally flawed, because of a failure to address critical issues pertaining to liability and injury compensation. There are three distinct aspects of the smallpox vaccination program injury debate – the concerns of the vaccine manufacturers, the vaccine administrators, and those exposed to the vaccinations themselves. Successful management of all three populations requires the timely development of appropriate legislative and regulatory protections.

Vaccine manufacturers, fearing sizeable litigation and liability costs which might arise out of the use of what is commonly seen as the most dangerous vaccine on the market, would not agree to manufacture and deliver the new smallpox vaccine doses without the government providing them with strong liability protection. Consequently, Congress, as part of the Homeland Security Act of 2002, included a provision which would shield pharmaceutical companies from liability for injuries arising out of the smallpox countermeasure manufacturing process [23]. Any claims related to countermeasure manufacture would be addressed through a federal compensation program, in which the United States would serve as the defendant. This provision also purported to protect those who administer smallpox vaccines but, as will be discussed below, the protections outlined within the law for that population were deemed to be unreasonably vague and ineffective in protecting all those involved in the vaccine distribution and administration process.

History demonstrates that the countermeasure manufacturers' fear of litigation is not misplaced. The most visible manifestation of such a concern – and the challenges which arise when attempting to rapidly develop and deploy a nationwide preparedness campaign in the face of an uncertain threat – arose during what came to be known as the Swine Flu Affair [24, 25]. On Friday, February 13, 1976, several soldiers training at Fort Dix, New Jersey, were reported to have fallen ill with influenza, with one of the recruits dying from the illness. Preliminary studies indicated that several of those who had the flu symptoms seemed to have a similar virus strain to that which caused the Great Influenza Pandemic of 1918, which was responsible for 40 million deaths worldwide, including half a million in the US. Media coverage of these events was widespread and stoked fears of a similar catastrophe. There was a public outcry for access to vaccinations. The public health community, in association with the pharmaceutical industry, felt it could rise to the occasion and rapidly produce and distribute enough vaccines to meet the demand. The CDC, headed by director David Sencer and advised by a panel of immunization experts, recommended to President Ford that a mass immunization campaign take place. President Ford announced that every man, woman and child should be immunized against the threat of swine flu, and Congress quickly passed a $134 million appropriation to pay for the effort.

The feeling was that it was better to be safe than sorry – that getting everyone vaccinated, even if the risk to the entire population was vague or overstated, was the most prudent approach to the potential outbreak. As one of the policymakers involved with the process indicated, there was also a dose of arrogance, heroic aspirations, and envy on the part of the public health policymakers when coming to this decision. As Reuel Stallones of the University of Texas School of Public Health (one of the members of the CDC advisory panel) later wrote:

> It was an opportunity to strike a blow for epidemiology in the
> interest of humanity. The rewards have gone overwhelmingly to
> molecular biology which doesn't do much for humanity.
> Epidemiology ranks low in the hierarchy – in the pecking order,
> the rewards system. Yet it holds the key to reducing lots of human
> suffering [26].

However, two large, interconnected roadblocks to implementing this policy were the issues of supply and liability. The pharmaceutical manufacturers' insurance companies stated that they would not cover injuries arising out of the manufacture and use of the swine flu vaccine, which in turn jeopardized the supply of vaccine.

Congress responded by passing legislation establishing the National Swine Flu Immunization Program of 1976 (the Swine Flu Act) [27], which modified the Federal Tort Claims Act to enable liability for problems arising out of vaccination manufacture to be underwritten by the government.

The Swine Flu vaccination program began on October 1, 1976, and free immunizations were made available to the public. In the weeks which followed, three elderly people in Pittsburgh died soon after receiving the vaccine. The media began to engage in "body count" reporting. The President and his family, in an attempt to allay public fears, went on prime-time television to receive vaccinations. In November, reports started coming in from several states of cases of Guillain–Barré syndrome, a paralytic disease which seemed to be connected with the vaccination. By this time, over 40 million people had received the shots. In December 1976, epidemiologists were able to more clearly draw a connection between the shot and the illness. By December 16, the program was concluded. More than one thousand of those who received shots claimed the vaccine caused them to contract Guillain–Barré syndrome, although, in hindsight, it seems many of these claims were dubious [28]. Because the Swine Flu Act's no-fault injury compensation program offered broad grounds on which a vaccine-injury claim could be filed, and the program did not limit the amount of damages which could be awarded [28], the government over the following decade paid more than $90 million in damages for such claims [29]. In the 1980 s, in response to concerns about childhood vaccine shortages, Congress granted similar protection to manufacturers of those vaccines by passing the National Childhood Vaccine Injury Act of 1986 [30]; the remedies available to those found to be injured by such vaccines were, however, far more limited than those found under the Swine Flu Act [28].

While the Homeland Security Act offered adequate protection to vaccine manufacturers, significant concerns remained about the level of protection the law afforded to those who would be responsible for implementation of the smallpox response plan. Because questions about vaccine-related injuries (both physical and economic) suffered by those administering, receiving, and exposed to those receiving, smallpox vaccines could not adequately be addressed before initiation of the smallpox vaccination program, hospitals, public-health departments, and healthcare workers were unwilling to voluntarily accept the risks associated with the smallpox vaccination in sufficient numbers to consider the smallpox vaccination program a success. It is, however, useful to examine the steps Congress and the administration took to try to alleviate such concerns, and how such efforts failed to overcome volunteer uncertainty.

Section 304 of the Homeland Security Act provided vaccine manufacturers, distributors, administrators, and officials, agents, or employees of such persons with protection from liability arising out of vaccinia-related injuries. This language, whereas broad, did not cover a broad enough population of people involved with the smallpox vaccination initiative. Ancillary participants in smallpox plans – such as those who might be involved in screening volunteers for contraindications – were not clearly covered under the statute. Section 304 also did not cover those who might be injured by the side-effects of vaccination, or those who might accidentally be injured by cases of contact vaccinia, in the absence of a negligent or wrongful act on the part of the vaccine manufacturer or administrator. Those who might find the need to seek redress for vaccine-related injuries were given little guidance from federal authorities and left with little legal recourse [31].

The vaccination program also created personnel management and public-relations challenges for hospitals. Hospitals expressed concern they might face substantial staffing shortages, because of vaccinia-related injuries to specialized health-care professionals and/or hospital employees and staff being required leaving work for up to 15 days to ensure protection against contact vaccinia cases spreading to vulnerable patients or other employees within the institution [32]. Although the Homeland Security Act offered hospitals adequate protection against lawsuits arising out of contact vaccinia cases (as long as hospital staff adhered to proper safety precautions), the negative economic impact of news of a contact vaccinia case occurring in a hospital could be devastating. Hospitals were also concerned about the economic impact of becoming designated as the local "smallpox hospital" in a national preparedness program [33]. While such a designation might reflect good citizenship in the short term, such a brand might reduce patient volume and negatively affect a hospital's long-term economic prospects.

The Secretary of Health and Human Services, sensitive to the liability concerns of health care providers and public-health departments arising out of the limited protections of Section 304, used the occasion of the late January 2003 declaration commencing the vaccination program to attempt to expand the smallpox liability protections available under the Homeland Security Act to include ancillary participants in the program [20]. Although no formal legal challenge was leveled against the Department of Health and Human Services for inclusion of this language, the declaration probably expanded the powers of the program beyond the scope of the enabling statute, and therefore did not ease doubts about injury protection held by possible program administrators. Neither Section 304 nor the January 2003 declaration offered assistance to the injured inoculated. The Secretary contended private health insurance and/or workers' compensation would cover vaccine-related injuries; there was, however, little evidence to support this assertion. Furthermore, such compensation programs would not cover instances of contact vaccinia contraction [28].

This continued ambiguity about whom and what would be covered under the federal compensation program raised the specter of considerable economic costs to be incurred by hospitals and public-health departments participating in the program, and curbed any enthusiasm they might have had for signing their organizations up to the effort [33, 34]. The lack of a clear, comprehensive compensation program also weighed heavily in many health professionals' conclusions that the personal risks they would assume by volunteering for the program outweighed their assessment of the value of the benefits which might be accrued to the population through their pre-event vaccination [35]. Concerns about the risks of the vaccination and the lack of injury compensation for volunteers entering the federal pre-event vaccination program, coupled with the uncertain benefits of participation in the program, led the IOM Committee to recommend that clear statements about these uncertainties be included as part of the vaccination in-formed consent process [8].

In April 2003, three months after Phase I of the smallpox vaccination program officially began, Congress finally passed the Smallpox Emergency Personnel Protection Act of 2003 (SEPPA) [35], a compensation program for those suffering debilitating side effects from receipt of the smallpox vaccine. This, however, was also seen as an inadequate incentive for hospitals, public-health departments, and health-care professionals to participate in the program, delivered far too late. Although this program, like the Swine Flu Act's compensation program, offered claimants a no-fault compensation system, and also expanded the number of people who would fall under the auspices of the federal compensation program, unlike the Swine Flu Act, there were strict limitations both on when injured parties could seek compensation from the government (they first had to investigate private insurance and workmen's compensation programs for reimbursement), and the amount of compensation an injured party could receive under the program. Furthermore, SEPPA was limited to compensation of health professionals partic-ipating in pre-event vaccination programs; health professionals and other emer-gency personnel who might be injured by receiving the smallpox vaccination after detection of a smallpox case would not be covered under the federal compensation program [31].

The failure of the smallpox vaccination plan offers useful insights into many of the legal and policy challenges faced in bioterrorism preparedness generally:

- attempting to guide national program planning while navigating through a splintered US public health system, with its 50 states with 50 separate sets of budgets, statutes, policy priorities, and personnel characteristics;
- economic and liability concerns of the businesses (hospitals, vaccine manufac-turers), agencies, and individuals relied upon to participate within the system;
- the setting of priorities on which preparedness efforts will be focused (e.g. single-agent planning or general preparedness efforts) [37]; and
- the challenges of making policy choices when faced with limited scientific and/ or national security information.

For example, similar concerns about agent-specific planning have been raised with regard to the Project BioShield Act of 2004 [38], which will devote $5.6 billion over 10 years toward the development and production of bioterrorist countermeasures, for example antibacterial and antiviral preparations and vaccines [39]. Although this will assist with the development of countermeasures for many dangerous agents, according to Ken Alibek and Charles Bailey the production of vaccines as a countermeasure for civilian populations may be misguided, because its effective-ness relies on the presence or absence of very specific circumstances, for example a known and limited target population for the vaccine and clear knowledge that the agent for which the vaccine is being created has not somehow been genetically altered to avoid available defense measures [39]. Furthermore, as the recent FDA recall of many pharmaceuticals indicates, enabling expedited peer review of bio-medical countermeasures raises concerns that avoidable side-effects may receive less-than-appropriate attention, because of the objectives of these new interventions. Finally, the use of bioterrorism preparedness funding has led to significant improvements in local, state and national surveillance efforts. This enhanced capacity has spawned innovative detection and surveillance initiatives, for example the national BioSense initiative [40], which will cull data from "clinical laboratories, hospital systems, ambulatory care sites, health plans, US Department of Defense and Veterans Administration medical treatment facilities, and pharmacy chains" to facilitate early detection of public health and bioterrorist-related outbreaks, and the National Bioterrorism Syndromic Surveillance Demonstration Program, which will cull data from health system and clinic electronic medical records to detect local bioterrorism outbreaks [41]. Although these programs may be significant strides toward improvement of the nation's bioterrorism response and preparedness, the potential usefulness of this new technology must be balanced against significant legal, ethical, and civil rights concerns, such as an individual's right to privacy.

Bibliography of Sources

K Alibek, C Bailey. BioShield or Biogap? *Biosecurity and Bioterrorism*, 2004, 2, 132–133.

G.J. Annas, Bioterrorism, Public Health, and Civil Liberties. *New Eng. J. Med.*, 2002, 346, 1337–1342.

G.J. Annas, Bioterrorism, Public Health, and Human Rights. *Health Aff.*, 2002, 21(6), 94–97.

J. Barbera, A. Macintyre, L. Gostin et al. Large-scale quarantine following biological terrorism in the United States, in Bioterrorism: Guidelines for Medical and Public Health Management. (D. A. Henderson, T. V. Inglesby, T. O'Toole, eds.) AMA Press: Chicago, 2002, 221–232, 222–223.

Z. Bashir, D. Brown, K. Dunkle, S. Kaba, C. McCarthy, The impact of federal funding on local bioterrorism preparedness. *J. Public Health Management Practice*, 2004, 10, 475–478.

R. Bayer and J. Colgrove, Public Health vs. Civil Liberties. *Science*, 2002, 297, 1811.

A.L. Benin, L. Dembry, E. D. Shapiro, E. S. Holmboe, Reasons physicians accepted or declined smallpox vaccine, February through April, 2003, *J. Gen. Intern. Med.*, 2004, 19, 85–89.

W.J. Bicknell, The case for voluntary smallpox vaccination. *N. Engl. J. Med.*, 2002, 346, 1323–1325.

D. Brown, A shot in the dark: Swine Flu's vaccine lessons. *Wash. Post*, May 27, 2002, at A9.

J.C. Butler, M. L. Cohen, C. R. Friedman, R. M. Scripp, C. G. Watz. Collaboration between public health and law enforcement: New paradigms and partnerships for bioterrorism planning and response. *Emerg. Infect. Dis.*, 2002, 8, 1152–1156.

H.W. Cohen, R. M. Gould, V. W. Sidel, The pitfalls of bioterrorism preparedness: The anthrax and smallpox experiences. *Am. J Public Health*, 2004, 94, 1667–1671.

C. Connolly, Smallpox vaccination for medical workers proposed, *Wash. Post*, September 4, 2002, p. A01.

C. Connolly, Smallpox plan may force other health cuts: States cite inability to fund vaccinations. *Wash. Post*, December 24, 2002, at A01.

Centers for Disease Control and Prevention, Smallpox Response Plan and Guidelines (Version 3.0), last updated November 26, 2002 (available at http://www.bt.cdc.gov/agent/smallpox/response-plan/index.asp).

Center for Law & the Public's Health at Georgetown and Johns Hopkins University, The Model State Emergency Health Powers Act Legislative Surveillance Table, available at http://www.publichealthlaw.net/MSEHPA/MSEHPA%20Surveillance.pdf.

Department of Health and Human Services, Declaration Regarding Administration of Smallpox Countermeasures, 68 Fed. Reg. 4,212 (Jan. 28, 2003).

V.S. Elliott, Public health funding: Feds giveth but the states taketh away. *American Medical News*, October 28, 2002 at 1.

Executive Order 13295, April 4, 2003 (available at http://www.fas.org/irp/offdocs/eo/eo-13295.htm).

Executive Order: Amendment to E.O. 13295 Relating to Certain Influenza Viruses and Quarantinable Communicable Diseases, April 1, 2005 (available at http://www.whitehouse.gov/news/releases/2005/04/20050401–6.html).

A.S. Fauci, Smallpox vaccination policy – the need for dialogue, *N. Engl. J. Med.*, 2002, 346, 1319–1320.

B. Gellman, 4 nations thought to possess smallpox. *Wash. Post*, November 5, 2002, at A1.

R.A. Goodman, J. W. Munson, K. Dammers, Z. Lazzarini, J. P. Barkley, Forensic epidemiology: law at the intersection of public health and criminal investigations. *J. Law Med. Ethics*, 2003, 31, 684–700.

L.O. Gostin, Public Health Law: Power, Duty, Restraint, (University of California Press: Berkeley and Los Angeles, 2000.

L.O. Gostin, Model State Emergency Health Powers Act, Draft #1: October 23, 2001 (on file with author).

L.O. Gostin, Model State Emergency Health Powers Act, Draft #2: December 21, 2001 (available at http://www.publichealthlaw.net).

L.O. Gostin, Public health law in an age of terrorism: Rethinking individual rights and common goods, *Health Aff.* 2002, 21(6) 79–93.

F.P. Grad, The Public Health Law Manual, 3rd Edition. American Public Health Association, Washington, 2005.

M. Greenberger, The 800 pound gorilla sleeps: the federal government's lackadaisical liability and compensation policies in the context of pre-event vaccine immunization programs, *J. Health Care Law & Policy*, 2005, 8, 7–37.

E. Gursky and A. Parikh, Lessons learned from the Phase I civilian smallpox program, *J. Health Care Law & Policy*, 2005, 8, 162–184, at 179 (citations omitted).

R.M. Henig, The people's health: A memoir of public health and its evolution at Harvard. (Joseph Henry Press, 1996), pp. 121–122.

J. G. Hodge, Jr., Bioterrorism law and policy: critical choices in public health, *J. Law, Med. & Ethics*, 2002, 30, 254–261.

R.E. Hoffman, Preparing for a bioterrorist attack: Legal and administrative strategies, *Emerg. Infect. Dis.*, 2003, 9, 241–245.

Homeland Security Act of 2002, Public Law No.107–296.

http://www.cdc.gov/od/oc/media/spvaccin.htm.

Institute of Medicine, Committee on Smallpox Vaccination Program Implementation, Review of the Centers for Disease Control and Prevention's Smallpox Vaccination Program Implementation: Letter Report #1 (hereinafter IOM Letter Report #1), January 17, 2003 (available at http://www.iom.edu/project.asp?id=4781).

Institute of Medicine, Committee on Smallpox Vaccination Program Implementation, Review of the Centers for Disease Control and Prevention's Smallpox Vaccination Program Implementation: Letter Report #2 (hereinafter IOM Letter Report #2), March 27, 2003 (available at http://www.iom.edu/project.asp?id=4781).

Institute of Medicine, Committee on Smallpox Vaccination Program Implementation, Review of the Centers for Disease Control and Prevention's Smallpox Vaccination Program Implementation: Letter Report #3 (hereinafter IOM Letter Report #3), May 27, 2003 (available at http://www.iom.edu/project.asp?id=4781).

Institute of Medicine, Committee on Smallpox Vaccination Program Implementation, Review of the Centers for Disease Control and Prevention's Smallpox Vaccination Program Implementation: Letter Report #4 (hereinafter IOM Letter Report #4), August 12, 2003 (available at http://www.iom.edu/project.asp?id=4781).

Institute of Medicine, Committee on Smallpox Vaccination Program Implementation, Review of the Centers for Disease Control and Prevention's Smallpox Vaccination Program Implementation: Letter Report #5 (hereinafter IOM Letter Report #5), December 19, 2003 (available at http://www.iom.edu/project.asp?id=4781).

Institute of Medicine, Committee on Smallpox Vaccination Program Implementation, Review of the Centers for Disease Control and Prevention's Smallpox Vaccination Program Implementation: Letter Report #6 (hereinafter IOM Letter Report #6), July 6, 2004 (available at http://www.iom.edu/project.asp?id=4781).

A.R. Kemper, A. E. Cowan, P. L. Y.H. Ching, M. M. Davis, E. J. Kennedy, S. J. Clark, G. L. Freed, Hospital decision-making regarding the smallpox pre-event vaccination program. *Biosecurity and Bioterrorism*, 2005, 3, 23–30.

J.W. Loonsk, BioSense – a national initiative for early detection and quantification of public health emergencies. *MMWR Morb Mortal Wkly Rep.* 2004, *53 Suppl*, 53–55.

W.K. Mariner, G. J. Annas, L. H. Glantz. Jacobson v Massachusetts: it's not your great-great-grandfather's public health law. *Am. J. Public Health*, 2005, *95*, 581–590.

T. May, R. D. Silverman, Should smallpox vaccine be made available to the general public? *Kennedy Inst. Ethics J.*, 2003, 13(2), 67–82.

T. May, R. Silverman. Bioterrorism defense priorities. *Science*, 2003, *301*, 17.

D.G. McNeil, Jr., 2 Programs To Vaccinate For Smallpox Come to a Halt. *New York Times*, June 19, 2003, at A13.

A.D. Moulton, R. N. Gottfried, R. A. Goodman, A. M. Murphy, R. D. Rawson, What is public health legal preparedness? *J. Law Med. Ethics*, 2003, 31, 672–683.

National association of county and city health officials, Research brief #9: Impact of smallpox vaccination program on local public health services, February 2003 (available at http://archive.naccho.org/Documents/Research_Brief_9.pdf).

National partnership for immunization, ACIP Meeting – Smallpox Vaccination Recommendations (June 19–20, 2002). (available at http://www.partnersforimmunization.org/meetingupdates_aicp.html).

National swine flu immunization program of 1976, Pub. L. No. 94–380, 90 Stat. 1113 (codified as amended at 42 U. S. C.A. § 247 b).

National vaccine injury act of 1986, Pub. L. No. 99–660, 100 Stat. 3756 (1986) (codified at 42 U. S. C.A. § 300aa-1 to 300aa-33).

R.E. Neustadt, H. V. Feinberg, J. A. Califano, Jr., The swine flu affair: Decision-making on a slippery disease. (U. S. Department of Health, Education and Welfare, 1978).

Office of the President of the United States, Protecting Americans: Smallpox Vaccination Program, (available at http://www.whitehouse.gov/news/releases/2002/12/20021213–1.html).

W.E. Parmet, R. A. Goodman, A. Farber. Individual rights versus the public's health – 100 years after Jacobson v. Massachusetts. *N. Engl. J. Med.* 2005, *352*, 652–654.

Public Health Security and Bioterrorism Preparedness and Response Act, Public Law 107–188.

R. Pear, States are facing big fiscal crises, governor's report. *N. Y. Times*, November 26, 2002, at A1.

R. Preston, The demon in the freezer. (New York: Random House, 2002).

E.P. Richards, Collaboration between public health and law enforcement: the Constitutional challenge, *Emerg. Infect. Dis.*, 2002, 8, 1157–1159.

E.P. Richards and K. C. Rathbun, Legislative alternatives to the model state emergency health powers act (MSEHPA): LSU program in law, science, and public health white paper #2, April 21, 2003, available at http://biotech.law.lsu.edu/blaw/bt/MSEHPA_review.htm.

E.P. Richards, K. C. Rathbun, J. Gold, The smallpox vaccination campaign of 2003: Why did it fail and what are the lessons for bioterrorism preparedness? *Louisiana Law Rev.*, 2004, 64, 851–904.

A. Schuler, Billions for biodefense: Federal agency biodefense funding, FY20001–FY2005, *Biosecurity & Bioterrorism*, 2004, 2(2):86–96.

A.M. Silverstein, Pure politics and impure science: The Swine Flu affair (Johns Hopkins University Press, 1981)

Smallpox Emergency Personnel Protection Act of 2003, Pub. L. No. 108–20, 117 Stat 638 (2003) (codified as amended at 42 U. S. C.A. §§ 233, 239 & 239a–h).

P.F. Smith, H. G. Chang, K. A. Sepkowitz. Inpatients at risk of contact vaccinia from immunized health care workers. *JAMA*, 2003, *289*, 1512–1513.

A.B. Staiti, A. Katz, J. F. Hoadley, Has bioterrorism preparedness improved public health? *Cent. Stud. Health Syst. Change*, Issue Brief, July 2003, 65, 1–4.

Turning Point Program, available at http://turningpointprogram.org/.

S.J. Wilson, Factors affecting implementation of the U. S. smallpox vaccination program, 2003, *Pub. Health Reports*, 2005, 120, 3–5.

K.R. Wing, The Law and the Public's Health, 6th Edition. Health Administration Press, Chicago, 2003.

W.K. Yih, B. Caldwell, R. Harmon et al. National Bioterrorism Syndromic Surveillance Demonstration Program. *MMWR Morb Mortal Wkly Rep.* 2004, *53 Suppl*, 43–49.

References

1 B Gellman, *4 nations thought to possess smallpox*, Wash Post, November 5, 2002, at A1.

2 Richard Preston, The Demon In The Freezer (Random House 2002)

3 AS Fauci, *Smallpox vaccination policy – the need for dialogue*, N Engl J Med, 2002, *346*, 1319–1320

4 WJ Bicknell, *The case for voluntary smallpox vaccination.* N Engl J Med. 2002, *346*, 1323–1325.

5 T May, RD Silverman, *Should smallpox vaccine be made available to the general public?* Kennedy Inst Ethics J. 2003, *13(2)*, 67–82.

6 National Partnership for Immunization, ACIP Meeting – Smallpox Vaccination Recommendations (June 19–20, 2002). (available at http://www.partnersforimmunization.org/meetingupdates_aicp.html)

7 C Connolly, Smallpox Vaccination for Medical Workers Proposed, Wash Post, September 4, 2002, p. A01.

8 Institute of Medicine, Committee on Smallpox Vaccination Program Implementation, Review of the Centers for Disease Control and Prevention's Smallpox Vaccination Program Implementation: Letter Report #1 (hereinafter IOM Letter Report #1), January 17, 2003 (available at http://www.iom.edu/project.asp?id=4781)

9 Institute of Medicine, Committee on Smallpox Vaccination Program Implementation, Review of the Centers for Disease Control and Prevention's Smallpox Vaccination Program Implementation: Letter Report #2 (hereinafter IOM Letter Report #2), March 27, 2003 (available at http://www.iom.edu/project.asp?id=4781)

10 Institute of Medicine, Committee on Smallpox Vaccination Program Implementation, Review of the Centers for Disease Control and Prevention's Smallpox Vaccination Program Implementation: Letter Report #3 (hereinafter IOM Letter Report #3), May 27, 2003 (available at http://www.iom.edu/project.asp?id=4781)

11 Institute of Medicine, Committee on Smallpox Vaccination Program Implementation, Review of the Centers for Disease Control and Prevention's Smallpox Vaccination Program Implementation: Letter Report #4 (hereinafter IOM Letter Report #4), August 12, 2003 (available at http://www.iom.edu/project.asp?id=4781)

12 Institute of Medicine, Committee on Smallpox Vaccination Program Implementation, Review of the Centers for Disease Control and Prevention's Smallpox Vaccination Program Implementation: Letter Report #5 (hereinafter IOM Letter Report #5), December 19, 2003 (available at http://www.iom.edu/project.asp?id=4781)

13 Institute of Medicine, Committee on Smallpox Vaccination Program Implementation, Review of the Centers for Disease Control and Prevention's Smallpox Vaccination Program Implementation: Letter Report #6 (hereinafter IOM Letter Report #6), July 6, 2004 (available at http://www.iom.edu/project.asp?id=4781).

14 http://www.whitehouse.gov/news/releases/2002/12/20021213−1.html

15 LA Altman & A O'Connor, *Health Officials Fear Local Impact Of Smallpox Plan*, N. Y. Times, January 5, 2003, Sec 1, pg. 1.

16 C Connolly, *Smallpox Plan May Force Other Health Cuts: States Cite Inability To Fund Vaccinations*, Wash Post, December 24, 2002, at A01.

17 National Association of County and City Health Officials, *Research Brief #9: Impact of Smallpox Vaccination Program on Local Public Health Services*, February 2003 (available at http://archive.naccho.org/Documents/Research_Brief_9.pdf)

18 E Gursky and A Parikh, *Lessons learned from the Phase I civilian smallpox program*, J Health Care Law & Policy, 2005, *8*, 162–184, at 179 (citations omitted)

19 M.A. J. McKenna, *Most metro hospitals buck smallpox plan*, Atlanta Journal-Constitution, January 23, 2003, pg. 3A

20 Department of Health and Human Services, Declaration Regarding Administration of Smallpox Countermeasures, 68 Fed. Reg. 4,212 (Jan. 28, 2003).

21 DG McNeil, Jr., *2 Programs To Vaccinate For Smallpox Come to a Halt*, New York Times, June 19, 2003, at A13.

22 http://www.cdc.gov/od/oc/media/spvaccin.htm

23 Homeland Security Act of 2002, Public Law 107−296, Sec. 304.

24 RE Neustadt, HV Feinberg, JA Califano, Jr. The Swine Flu Affair: Decision − Making on a Slippery Disease (U. S. Department of Health, Education and Welfare, 1978).

25 AM Silverstein, Pure Politics and Impure Science: The Swine Flu Affair (Johns Hopkins University Press, 1981)

26 RM Henig, The People's Health: A Memoir of Public Health and Its Evolution at Harvard (Joseph Henry Press, 1996), pp. 121–122

27 National Swine Flu Immunization Program of 1976, Pub. L. No. 94−380, 90 Stat. 1113 (codified as amended at 42 U. S. C.A. § 247 b)

28 M Greenberger, *The 800 pound gorilla sleeps: the federal government's lackadaisical liability and compensation policies in the context of pre-event vaccine immunization programs*, J Health Care Law & Policy, 2005, *8*, 7–37.

29 D Brown, *A shot in the dark: Swine Flu's vaccine lessons*, Wash. Post, May 27, 2002, at A9.

30 National Vaccine Injury Act of 1986, Pub. L. No. 99−660, 100 Stat. 3756 **1986** (codified at 42 U. S. C.A. § 300aa-1 to 300aa-33).

31 EP Richards, KC Rathbun, J Gold, *The smallpox vaccination campaign of 2003: why did it fail and what are the lessons for bioterrorism preparedness?* Louisiana Law Rev., 2004, *64*, 851–904

32 PF Smith, HG Chang, KA Sepkowitz. Inpatients at risk of contact vaccinia from immunized health care workers. JAMA. 2003, *289*, 1512–1513.

33 AR Kemper, AE Cowan, PLYH Ching, MM Davis, EJ Kennedy, SJ Clark, GL Freed, *Hospital decision-making regarding the smallpox pre-event vaccination program*, Biosecurity and Bioterrorism, 2005, *3*, 23–30.

34 SJ Wilson, *Factors affecting implementation of the U. S. smallpox vaccination program, 2003*, Pub Health Reports, 2005, *120*, 3–5.

35 AL Benin, L Dembry, ED Shapiro, ES Holmboe, *Reasons physicians accepted or declined smallpox vaccine, February through April, 2003*, J Gen Intern Med., 2004, *19*, 85–89.

36 Smallpox Emergency Personnel Protection Act of 2003, Pub. L. No. 108–20, 117 Stat 638 **2003** (codified as amended at 42 U. S. C.A. §§ 233, 239 & 239a-h).

37 T May, R Silverman. Bioterrorism defense priorities. *Science*, 2003, *301*, 17

38 Project BioShield Act of 2004, Pub. Law No. 108–276.

39 K Alibek, C Bailey. BioShield or Biogap? *Biosecurity and Bioterrorism*, 2004, *2*, 132–133.

40 J.W. Loonsk, BioSense – a national initiative for early detection and quantification of public health emergencies. *MMWR Morb Mortal Wkly Rep.* 2004, *53 Suppl*, 53–55.

41 WK Yih, B Caldwell, R Harmon et al. National Bioterrorism Syndromic Surveillance Demonstration Program. *MMWR Morb Mortal Wkly Rep.* 2004, *53 Suppl*, 43–49.

11

Legal Preparedness: The Modernization of State, National, and International Public Health Law

Ross D. Silverman

11.1
Legal Preparedness: Sources of Power and Limits

Public health actions may be guided in part by federal and/or state constitutions, acts passed by the federal and/or state legislatures, regulations passed by state or federal agencies, executive orders of the President or state governor, interstate compacts, or local ordinances. As is discussed in greater depth below, most public-health activity and legal authority is based in the states; on issues pertaining to bioterrorism, however, the federal government is very influential in shaping local public health practices, priorities, and policies.

In general, public-health powers are defined by a legislature in broad enabling statutes, leaving the finer points of how best to address a public health concern to the public-health agency's expertise. For example, in 2002 Congress passed the Public Health Security and Bioterrorism Preparedness and Response Act (also known as The Bioterrorism Act of 2002) [1]. Through this far-reaching legislation, Congress, among other measures:

1. created an Assistant Secretary for Public-health emergency Preparedness within the Department of Health and Human Services;
2. allocated hundreds of millions of dollars toward bolstering the strategic national stockpile, the federal cache of vaccines, antibiotics, antitoxins, and other medical countermeasures for biological, radiological, and chemical agents;
3. authorized the Centers for Disease Control and Prevention to distribute billions of federal dollars through grants to state and local public-health agencies for bioterrorism and emerging infectious disease preparedness;
4. authorized more than $500 million of grant funding to be distributed through the Health Resources and Services Administration to assist hospitals and other health-care facilities with bioterrorism and mass casualty response capacity;

Bioterrorism Preparedness. Edited by Nancy Khardori
Copyright © 2006 WILEY-VCH Verlag GmbH & Co. KGaA, Weinheim
ISBN: 3-527-31235-8

5. required registration of laboratories and individuals working with those select agents deemed to pose significant public health and safety threats, and the creation of regulations to bolster the safety of handling such agents; and

6. instituted measures to bolster the protection of the nation's food, drug, and water supplies.

Another example is the Homeland Security Act of 2002 [2], through which Congress created the federal government's largest agency, and authorized it to coordinate the nation's national security, including emergency preparedness and response and the development and promotion of research and technology to combat biological, chemical, radiological, and nuclear threats.

States are usually seen as the primary public-health authorities, because they have been granted very broad constitutional powers, known as the police powers, to protect the public health, welfare, and safety of the community [3]. In the context of bioterrorism preparedness, these powers include the power to enforce isolation and quarantine, conduct surveillance and laboratory testing, compel examination or treatment, seize, confiscate, or destroy goods, and license health-care providers [4, 5]. Although such conduct may infringe individual rights, the power to protect the public's health may supersede such liberties in cases of significant threats to the greater population. This power has two primary sources, one legal and one moral.

An action taken by a public-health authority against an individual or his property may be challenged in court. The most important case through which to assess the scope of public health law powers is the 1905 case of *Jacobson v. Massachusetts* [6]. In this case the Supreme Court determined that a local public-health agency had the constitutional authority to compel an individual either to receive a smallpox vaccination or pay a $5 fine, when the agency had determined that the community was at risk of a smallpox outbreak. In response to the defendant's complaint of impingement upon his individual rights, Justice Harlan, writing for the court, stated as follows:

The defendant insists that his liberty is invaded when the State subjects him to fine or imprisonment for neglecting or refusing to submit to vaccination; that a compulsory vaccination law is unreasonable, arbitrary and oppressive, and, therefore, hostile to the inherent right of every freeman to care for his own body and health in such way as to him seems best; and that the execution of such a law against one who objects to vaccination, no matter for what reason, is nothing short of an assault upon his person. But the liberty secured by the Constitution of the United States to every person within its jurisdiction does not import an absolute right in each person to be, at all times and in all circumstances, wholly freed from restraint. There are manifold restraints to which every person is necessarily subject for the common good. On any other basis organized society could not exist with safety to its members [6].

This passage alludes to the moral authority which buttresses public-health power at the state level also. This moral authority has been described as a "public-health contract" between the state and members of society, in which "individuals agree to forgo certain rights and liberties, if necessary, to prevent a significant risk to other persons. Civil rights and liberties are subject to limitation because each person gains the benefits of living in a healthier and safer society." [7].

Although this case and the public health contract do give the public-health authority broad powers, these powers are not unbounded. Courts should assess public health actions on the grounds of necessity, reasonable means, proportionality, and harm-avoidance [8]. Over the course of the past 100 years increased attention and importance has been given by the courts and society to individual and human rights, and many scholars feel that, were the facts of the Jacobson case heard today, the outcome might be very different [9, 10]. Insofar as bioterrorism preparedness is concerned, however, and in the absence of evidence of malicious intent on the part of the state, history has shown that courts are loathe to challenge the expertise of a public-health authority on matters in which the state has determined there is a significant, immediate threat to the public health, welfare, and safety.

One of the most important public-health powers during a bioterrorism event is the initiation of an action of quarantine or isolation, which is one of the oldest manifestations of state public-health authority. The process of initiating such an action varies from state to state. In general, a quarantine action begins with an order which, depending on the jurisdiction, may come directly from the public-health authority, or may require a judge to issue a court order to execute an action to detain a contagious (or suspected-to-be-contagious) individual.

State "police powers" should not be confused with state and federal criminal law enforcement powers. Because bioterrorism is both a public health concern and a criminal act, however, response efforts are likely to include both criminal justice and public-health authorities. These two disciplines have distinct procedures, hierarchies, objectives, and languages, and dramatically different constitutional limits on their authority [11–13]. Those planning bioterrorism-response efforts must work collaboratively to determine an approach which will maximize the efficiency and effectiveness of such interactions.

11.2
Federal Public-health Authority

While local and state authorities are the traditional domain in which to address public health concerns and draft public-health policy, there are many reasons to justify increased Federal government influence on, and responsibility for, addressing critical public health concerns such as bioterrorism and emerging infectious diseases. The federal government can draw on greater financial resources and has far more financial flexibility than states and localities. For example, the federal government may undertake deficit spending in times of crisis, freeing up additional funds for initiatives such as public-health infrastructure funding, whereas most state governments have faced critical fiscal crises for several years [14], have a finite pool of funding upon which to draw, and are mandated by state constitutions to balance their annual budget. This means that public-health concerns not only face the prospect of being under-funded generally by state legislatures, but may also be constrained by the intense competition for funds they face from other state budgeting responsibilities with consistently rising costs and a perceived more "immediate" need (e.g. education, Medicaid). Significant increases in federal funding allocated for public-health infrastructure improvements and bioterrorism preparedness efforts [15, 16] (discussed further below) may have presented an opportunity for anxious state legislatures to make sizeable and harmful cuts to public-health funding [17], thereby undermining any structural improvements made to the public health system generally, and preparedness efforts specifically.

Other arguments in favor of federal, rather than state and local, leadership on bioterrorism law and policy include:

1. that bioterrorism is an issue not only of public health, but of national security, a wholly federal responsibility (in fact, the legislation establishing the Department of Homeland Security [2] authorized the agency to direct a number of public health-related activities previously managed by the Department of Health and Human Services);

2. the interstate and international dangers posed by bioterrorism, which would probably outstrip any local or state jurisdiction and exploit gaps in a state-based public-health legal system;

3. the power of the federal government to use its stronger position in the marketplace to negotiate prices for services and countermeasures and to address issues such as countermeasure safety and patents; and

4. the institutional experience found in federal agencies concerning emerging and rare infectious disease identification, treatment, and surveillance, and emergency response management, would enable more rapid and consistent response to an incident in any particular location nationwide [18].

In addition to providing research and political expertise on public-health preparedness issues, Congress and federal agencies such as the Centers for Disease Control and Prevention and the Department of Homeland Security use their spending authority to shape laws and policies related to preparedness. How the federal government wields its "power of the purse" is an important driver of state and local bioterrorism defense policy. Even in circumstances in which the federal government defers to state and local policymakers in how to structure bioterrorism-response initiatives, the contingencies the federal government may attach to the receipt of this significant federal funding can have a dramatic impact on the priorities and policies local public-health authorities ultimately adopt. During fiscal years 2001 to 2004 approximately $14.5 billion was allocated by the federal government toward funding civilian bioterrorism-preparedness measures, and many of these dollars contained specific policy requirements to obtain funding [19].

The creation of the Department of Homeland Security, and the billions of dollars it has distributed toward public-health preparedness efforts, has led to the reshaping of local of public-health agency priorities, policies, and infrastructure [20]. One of the most significant examples of the Federal government using its "power of the purse" to effect change in state and local public-health policy is the coordination of national, state, and local emergency response planning. On February 28, 2003, the President issued a Homeland Security Presidential Directive (HSPD-5), which directed the Secretary of Homeland Security to create and oversee a national incident-management system (NIMS) and create a national-response plan (NRP). Through the NIMS and NRP, the federal government could lead efforts to improve coordination between local, state, and federal emergency response agencies and provide a framework for more streamlined incident command and response. Federal agencies now require that those seeking federal preparedness grants, contracts, or other funding (including bioterrorism funding) adopt the NIMS guidelines. A full discussion of the NIMS guidelines is beyond the scope of this work; the NIMS document is, however, available online at http://www.nimsonline.com/nims_3_04/index.htm, with additional information about the program available at the web sites of the Federal Emergency Management Agency (http://www.fema.gov), and NIMSonline.com.

11.3
Federal Isolation and Quarantine Powers

Although states are the front-line agents of enforcement of quarantine and isolation actions, the Federal government maintains authority to use such powers to intercede in cases of detected and suspected interstate or international infectious disease threats. The Federal Public Health Service Act gives the Secretary of the Department of Health and Human Services the authority to create and enforce regulations needed "to prevent the introduction, transmission, or spread of communicable diseases from foreign countries into the states or possessions, or from one state or possession into any other state or possession." [21]. Federal regulations vest this power in the Director of the Centers for Disease Control and Prevention [22]. State law takes precedent in most quarantine and isolation circumstances unless state actions are found to conflict with federal action [23]. As described in the Federal quarantine regulations, this occurs:

Whenever the Director of the Centers for Disease Control and Prevention determines that the measures taken by health authorities of any state or possession (including political subdivisions thereof) are insufficient to prevent the spread of any of the communicable diseases from such state or possession to any other state or possession [24].

Executive orders establish agents against which federal quarantine may be used. These executive orders are made on the recommendation of the Secretary of Health and Human Services and in consultation with the Surgeon General [25]. The current list of communicable diseases for which the Centers of Disease Control and Prevention may institute a quarantine action are cholera, diphtheria, infectious tuberculosis, plague, smallpox, yellow fever, viral hemorrhagic fevers, severe acute respiratory syndrome (SARS) [26], and avian influenza [27].

The highly mobile population of the United States, and the inconsistency of public health laws from one state to another, make the Federal power to address interstate communicable disease threats critically important. Federal quarantine laws and regulations give the Department of Health and Human Services authorization to apprehend and examine "any individual reasonably believed to be infected with a communicable disease in a qualifying stage and (A) to be moving or about to move from a state to another state; or (B) to be a probable source of infection to individuals who, while infected with such disease in a qualifying stage, will be moving from a state to another state" [28]. The "qualifying stage" language enables preemptive action by the Federal government to intervene if they suspect an individual is infected with a disease likely to cause a public-health emergency if transmitted to others which is either communicable or in a precommunicable state [29]. Another critical component of Federal quarantine authority is the right to place prohibitions on interstate travel [30].

During a time of war the power of federal government increases, and individuals reasonably believed to be infected with a disease specified by executive order, and a source of infection to members of the armed forces, may be apprehended and examined by federal authorities. These powers are also extended to address similar threats to those who produce or transport supplies for the military [31].

11.4
International Health Regulations

In May 2005, the World Health Assembly adopted the revised International Health Regulations (IHR) [32], a significant new international tool in the fight against the spread of global public-health threats, including those of a bioterrorist nature. These regulations will go into effect in 2007. The emergence of many novel diseases, from HIV/AIDS to avian influenza, the re-emergence of old diseases, for example tuberculosis and polio, and the increasing level of global interconnectedness in travel and trade, were driving forces behind reconsideration of the World Health Organization's outdated and ineffectual authority. A January 2003 resolution of the Executive Board of the World Health Organization stated the three principles guiding the revision of the International Health Regulations:

1. the IHR should require that all types of public health risk which raise international concern be reported;

2. international trade and travel concerns should be respected, and stigmatization and unnecessary adverse effects on these areas should be avoided; and

3. surveillance and detection systems should be modified to address new and re-emerging infectious disease threats [33].

As these principles indicate, the new IHR seeks to build upon an improved capacity to share information and surveillance data worldwide.

Whereas the previous version of the International Health Regulations was limited to issues pertaining to cholera, plague, and yellow fever, the new IHR maximizes the flexibility of the World Health Organization to respond to public health threats affecting member states. This is best seen in the IHR's definition of a "public-health emergency of international concern", which the regulations define as "an extraordinary event which is determined … (i) to constitute a public health risk to other States through the international spread of disease and (ii) to potentially require a coordinated international response"[32]. This would include not only traditional public health concerns, for example emerging infectious diseases and bioterrorism, but also emergencies involving chemical, radiological, and nuclear agents.

Some public health threats, for example detection of a case of smallpox, poliomyelitis caused by wild-type poliovirus, human influenza caused by a new subtype, and severe acute respiratory syndrome (SARS), trigger a responsibility to immediately report the case to the World Health Organization. In all other public health cases an algorithm for determining whether a public-health emergency of international concern exists should be used when a triggering public health event has been detected. The algorithm includes addressing the questions:

1. Is the public health impact of the event serious?
2. Is the event unusual or unexpected?
3. Is there a significant risk of international spread? and
4. Is there a significant risk of international travel or trade restrictions [32, Annex 2]?

If it is determined that the incident must be reported to the WHO as a public-health emergency of international concern, the WHO may make recommendations to the member state on how to address the emergency. Although one of the underlying goals of the IHR is to minimize the impact on trade and travel, it would be within the power of the WHO to use the economic impact that a recommendation of restrictions on trade and travel to a particular country might create to influence cooperation on matters concerning international public health threats.

In addition to expanding the scope of WHO power, the IHR also serves as a mechanism to compel member states to improve their core capacity to address issues pertaining to public-health surveillance, reporting, notification, verification, response, and collaboration internally, and improve upon their ability to manage air, sea, and ground travel crossings.

Overall, the IHR holds the promise of improving the clarity, transparency, and certainty of international public-health emergency response. Continued collaboration through the WHO and other international organizations, for example the World Trade Organization, on issues of health will help to improve the international protection available to all nations in matters of infectious disease and bioterrorism control.

11.5
Legal Preparedness

Although state and federal public-health powers may be clear, it is important for public health leaders to question whether these powers have been successfully translated into laws, policies, and procedures which enable the effective protection of the public's health, and whether those who would create or utilize such powers are adequately equipped for their responsibilities. This critical process of evaluation and reform is known as "public health legal preparedness." Preparedness may be achieved by:

1. review and revision of public health legal authorities;
2. identification, assessment and improvement of the core competencies of those who develop, implement and interpret public-health authority (e.g. judges, legislators, public health professionals, etc.);
3. improvement of information available to assist public health agents in decision-making (e.g., practice guidelines); and
4. coordination of legal authorities horizontally (e.g. between states or local public-health agencies) and vertically (e.g. between local, state and federal authorities) in an effort to eliminate gaps in public health laws and improve interjurisdictional cooperation and response.

This enables more effective identification of public-health authority held by government entities, initiation of efforts to improve, organize and standardize state public health laws to address both ancient and modern public health concerns, and ensures that the application of public health laws adequately balance critical human rights and due process concerns with state public health protection demands [34]. For example, a state may undertake an examination of whether current laws inhibit effective distribution of medications and health-professional licensure in times of emergency, or whether current regulations pertaining to quarantine and isolation afford sufficient protection of individual liberty rights [35].

To aid in the evaluation of public-health regulations, Lawrence O. Gostin developed a five-step evaluation process through which public-health authorities should be required to justify public-health intervention. First, by scientific means public-health authorities must demonstrate the presence of a significant risk. Second, by use of the means/ends test, they must demonstrate that the intervention effectively reduces such risks. Third, by cost–benefit analysis, authorities should demonstrate that the benefits of the intervention outweigh the intervention's economic costs. Fourth, by examination of the invasiveness, frequency, and duration of the intervention, they should demonstrate that any human rights burdens placed on individuals are reasonable. Fifth, by comprehensive review of costs, benefits, and population needs, public-health authorities must demonstrate the fair distribution of public-health intervention throughout society [5].

The legal preparedness process has re-energized the field of public health law over the past two decades, and is a central component of current public-health preparedness efforts. Recent examples of public health legal preparedness initiatives include the TOPOFF (top officials), TOPOFF2, and TOPOFF3 bioterrorism response exercises, the work of the Turning-Point Initiative [36] sponsored by the Robert Wood Johnson and W. K. Kellogg Foundations, and the development of model acts, such as the Model State Emergency Health Powers Act, discussed in greater detail below.

11.6
Legal Preparedness in Action: The Model State Emergency Health Powers Act

Within six weeks of the September 11, 2001, attacks the Center for the Law and the Public's Health at Georgetown and Johns Hopkins Universities (the Center) released a Model State Emergency Health Powers Act (MSEHPA) [37]. The MSEH-PA was drafted at the request of the CDC to help state governments with their examination of the adequacy of state public-health powers to address public-health emergencies. According to the act's primary author, Lawrence O. Gostin:

MSEHA is structured to reflect five basic public-health functions to be facilitated by law: preparedness, surveillance, management of property, protection of persons, and public information and communication [38].

The act, which uses the expansive state police power authority described in *Jacobson v. Massachusetts* as a philosophical foundation, was touted as a way of modernizing state public-health laws, which in turn might clarify state authority, improve coordination of intrastate and interstate responses, and maintain an appropriate balance between individual rights and the power of the state police to protect the public's health, safety, and security [38].

After receiving significant feedback on the first draft of the act (see further discussion below), the Center released a second draft on December 21, 2001 [39]. This draft made many structural changes and refinements, including:

1. narrowing the definition of "public-health emergency," including eliminating from the text reference to epidemic and pandemic diseases, making optional language pertaining to response to natural disasters and chemical or nuclear attacks or incidents, and limiting the scope of an emergency to those circumstances which pose a high probability of morbidity, mortality, or serious or long-term disability;

2. clarification of powers surrounding public-health emergency declarations (creating a general requirement that such emergencies be declared in consultation with the state public-health authority), and response and conclusion; and
3. elimination of criminal penalties for an individual's failure to accept orders requiring the individual to administer or accept mandatory vaccination, testing, and treatment.

Two additional significant changes in the second draft of the MSEHPA are the creation of separate statutory definitions for isolation and quarantine actions, and the augmentation of due process rights for those subject to confinement. Isolation is defined as "the physical separation and confinement of an individual or groups of individuals *who are infected or reasonably believed to be infected* with a contagious or possibly contagious disease from non-isolated individuals, to prevent or limit the transmission of the disease to non-isolated individuals" [39, Sec.104 (h)] (emphasis added), whereas quarantine is the application of similar measures to those "who are or may have been exposed to a contagious or possibly contagious disease and who do not show signs or symptoms of a contagious disease [39, Sec.104 (o)]". Under the Act, isolation and quarantine may be applied to individuals or groups of individuals, and up to ten days of temporary isolation or quarantine may be imposed without notice "if delay in imposing the isolation or quarantine would significantly jeopardize the public-health authority's ability to prevent or limit the transmission of a contagious or possibly contagious disease to others." [39, Sec.104 (a)]. The public-health authority must apply these measures in the least restrictive means possible, leaving open the possibility of home confinement. Failure to adhere to an order for isolation or quarantine constitutes a criminal act.

Those facing isolation and confinement under the MSEHPA are entitled to a hearing before a trial court, in which the public-health authority must prove by a preponderance of the evidence that isolation or quarantine is reasonably necessary to address the public health threat. This hearing must generally be held within five days of the public-health authority's application to the court, although under extraordinary circumstances, the court proceedings may be deferred for up to ten additional days. Those who are ordered to isolation or quarantine are entitled to court-appointed counsel at the expense of the state, and may petition the trial court for release from confinement. Under the act, a release hearing must be held within five days of receipt of the request, or within twenty-four hours should the petition claim extraordinary circumstances justifying immediate release. The trial court may be in the district in which confinement is to occur or where the public-health emergency has been declared, or it may be a court designated by the state's public-health emergency plan.

Critiques of the MSEHPA within the public health law and ethics communities fell into three general categories:

1. civil liberties concerns;
2. questions about whether the act was necessary for the state to respond to a public health threat; and

3. suggestion that bioterrorism response is best handled by federal, rather than state, authorities.

Civil liberty concerns were widespread in the first draft of the act and were a primary driver behind the act's rapid revision [40, 41]. According to George J. Annas, the act's stringent quarantine and isolation requirements, use of extensive personal health information in surveillance and reporting, broad grant of authority to the state, and minimal accountability for instances of overzealous application of such authority gave the act the feel of a law "drafted for a different age; it is more appropriate for the United States of the 19th century than for the United States of the 21st century" [41]. By proposing such sweeping control over the public, health-care professionals, and entities and property, Annas and others argued that the act belied an inherent distrust in the public's willingness to cooperate in the face of a public-health emergency, worried that such "Draconian" measures might compromise public trust in the public health system and government authorities, and feared that civil rights and liberties might be easily and inappropriately overrun in the name of protecting the public [40, 41].

Other critics of the model act questioned whether states needed a model act at all [42]. Edward P. Richards and Katharine C. Rathbun argued that the broad police powers granted to state public-health agencies give more than ample authority to respond to bioterrorism threats, and that any state finding it necessary to clarify state authority could easily achieve this goal through the more flexible administrative law process than by attempting to overhaul state laws by adopting the model act. Furthermore, courts give great deference to public-health agency use of significant public-health powers in the face of significant public health threats, even if such actions impinge upon individual liberties, and attempts to delineate more specific procedures and due process protections might unduly burden public health threat response. According to the authors:

... judges will not stand in the way of emergency actions taken to protect the public from a clear and present danger, and if they do, the state appeals court will overturn their rulings in a matter of hours. From the colonial period until today, the history of judicial restraint on emergency powers is one of blind obedience to civil and military authority, not one of necessary actions thwarted by overly particular jurists. It is inconceivable that the courts would stand in the way of actions to control a major public health threat such as a smallpox outbreak, even if the state was clearly stepping beyond its statutory powers [42].

Finally, some scholars argued that federal, rather than state, authorities should manage bioterrorism response. Because bioterrorism concerns are a national and international security concern, reliance on local and state authorities might lead to significant gaps in public health protection and undue complexity and confusion in circumstances demanding clear direction and swift action [43].

Despite these concerns, the MSEHPA has received widespread support from state legislatures nationwide. As of July 1, 2004, 44 states had introduced the act or provisions of the act in bills or resolutions, and 33 states and the District of Columbia had passed bills or resolutions containing provisions from the model act or very similar language [44].

Bibliography of Sources

The bibliography of sources given in Chapter 10 is applicable to this chapter also.

References

1 Public Health Security and Bioterrorism Preparedness and Response Act, Public Law 107–188

2 Homeland Security Act of 2002, Public Law 107–296.

3 K.R. Wing, The Law and the Public's Health, 6th Edition. Health Administration Press, Chicago, 2003.

4 F.P. Grad, The Public Health Law Manual, 3rd Edition. American Public Health Association, Washington, 2005.

5 L.O. Gostin, Public Health Law: Power, Duty, Restraint, (University of California Press: Berkeley and Los Angeles, 2000.

6 Jacobson v. Massachusetts, 197 U. S. 11 **1905**.

7 J. Barbera, A. Macintyre, L. Gostin et al. *Large-scale quarantine following biological terrorism in the United States*, in Bioterrorism: Guidelines for Medical and Public Health Management (D. A. Henderson, T. V. Inglesby, T. O'Toole, eds.), AMA Press 2002, 221–232, 222–23..

8 LO Gostin, *Jacobson v Massachusetts at 100 years: police power and civil liberties in tension*. Am J Public Health. 2005, *95*, 576–581.

9 W.K. Mariner, G. J. Annas, L. H. Glantz. Jacobson v Massachusetts: it's not your great-great-grandfather's public health law. *Am. J. Public Health*, 2005, *95*, 581–590.

10 W.E. Parmet, R. A. Goodman, A. Farber. Individual rights versus the public's health – 100 years after Jacobson v. Massachusetts. *N. Engl. J. Med.* 2005, *352*, 652–654.

11 JC Butler, ML Cohen, CR Friedman, RM Scripp, CG Watz. Collaboration between public health and law enforcement: new paradigms and partnerships for bioterrorism planning and response. Emerg Infect Dis., 2002, 8, 1152–1156.

12 EP Richards, *Collaboration between public health and law enforcement: the Constitutional challenge*, Emerg Infect Dis., 2002, *8*, 1157–1159.

13 RA Goodman, JW Munson, K Dammers, Z Lazzarini, JP Barkley. *Forensic epidemiology: law at the intersection of public health and criminal investigations.* J Law Med Ethics, 2003, *31*, 684–700.

14 Robert Pear, *States are facing big fiscal crises, governors report*, NEW YORK TIMES, November 26, 2002, at A1.

15 Z. Bashir, D. Brown, K. Dunkle, S. Kaba, C. McCarthy, *The impact of federal funding on local bioterrorism preparedness*, J Public Health Management Practice, 2004, *10*, 475–478.

16 H.W. Cohen, R. M. Gould, V. W. Sidel, *The pitfalls of bioterrorism preparedness: the anthrax and smallpox experiences.* Am J Public Health, 2004, *94*, 1667–1671.

17 Victoria Stagg Elliott, *Public-health funding: Feds giveth but the states taketh away*, American Medical News, October 28, 2002 at 1.

18 J. G. Hodge, Jr. *Bioterrorism law and policy: critical choices in public health*, J Law, Med Ethics, 2002, *30*, 254–261.

19 A. Schuler, *Billions for Biodefense: Federal Agency Biodefense Funding, FY20001–FY2005*, Biosecurity Bioterrorism, 2004, 2(2):86–96.

20 AB Staiti, A Katz, JF Hoadley, *Has bioterrorism preparedness improved public health?* Issue Brief Cent Stud Health Syst Change, July 2003, *65*, 1–4.

21 42 U. S. C. § 264.

22 42 C. F. R. § 70.2.

23 42 U. S. C. 264(e).

24 42 C. F. R. 70.2.

25 42 U. S. C. § 264(b).

26 Executive Order 13295, April 4, 2003 (available at http://www.fas.org/irp/offdocs/eo/eo-13295.htm)

27 Executive Order: Amendment to E. O. 13295 Relating to Certain Influenza Viruses and Quarantinable Communicable Diseases, April 1, 2005 (available at http://www.whitehouse.gov/news/releases/2005/04/20050401–6.html)

28 42 U. S. C. 264(d).

29 42 U. S. C. 264(d)(2).

30 42 C. F. R. 70.3.

31 42 U. S. C. 266.

32 World Health Organization, International Health Regulations (available at http://www.who.int/csr/ihr/en/)

33 World Health Organization, Resolution of the Executive Board of the WHO: Revision of the International Health Regulations, EB111.R13, January 24 2003.

34 AD Moulton, RN Gottfried, RA Goodman, AM Murphy, RD Rawson, *What is public health legal preparedness?* J Law Med Ethics, 2003, *31*, 672–683.

35 R.E. Hoffman, *Preparing for a bioterrorist attack: legal and administrative strategies*, Emerg Infect Dis, 2003, *9*, 241–245.

36 Turning Point Program, http://turningpointprogram.org/

37 L.O. Gostin, Model State Emergency Health Powers Act, Draft #1: October 23, 2001 (on file with author)

38 L.O. Gostin, *Public health law in an age of terrorism: Rethinking individual rights and common goods*, Health Aff. 2002, *21(6)* 79–93.

39 L.O. Gostin, *Model State Emergency Health Powers Act, Draft #2: December 21, 2001* (available at http://www.publichealthlaw.net)

40 R. Bayer and J. Colgrove, *Public Health vs. Civil Liberties*, Science, 2002, *297*, 1811.

41 G.J. Annas, *Bioterrorism, Public Health, And Civil Liberties*, New Eng. J. Med., 2002, *346*, 1337–1342.

42 E.P. Richards and K. C. Rathbun, *Legislative Alternatives to the Model State Emergency Health Powers Act (MSEHPA): LSU Program in Law, Science, and Public Health White Paper #2*, April 21, 2003, available at http://biotech.-law.lsu.edu/blaw/bt/MSEHPA_review.htm.

43 G.J. Annas, *Bioterrorism, Public Health, And Human Rights*, Health Aff., 2002, *21(6)*, 94–97.

44 Center for Law and the Public's Health at Georgetown and Johns Hopkins University, *The Model State Emergency Health Powers Act Legislative Surveillance Table*, available at http://www.publichealthlaw.net/MSEHPA/ MSEHPA%20Surveillance.pdf.

Index

a

Abl tyrosine kinase   118
acetylcholin release   169
Actinobacillus   189
adult respiratory distress syndrome (ARDS)
   187, 211
Advisory Committee on Immunization
   Practices (ACIP)   116, 228
aeromedical isolation team   48
aerosol infectivity   10
aerosolization   4
agent
   – anticrop   4
   – antimicrobial agent   191
   – bacterial   14 ff.
   – contagious   9
   – incapacitating   4 ff.
   – lethal   9
   – microbial   9, 21
   – rickettsial   14 ff.
   – select   35
   – viral   22 f.
alphavirus   215
alphavirus encephalomyelitis   22
   – diagnosis   22
   – epidemiology   22
   – management   23
American Academy of Pediatrics' Task Force
   on Terrorism   74
American Society of Microbiology (ASM)
   172, 194
aminoglycoside   191
amoxicillin   17
amplified fragment polymorphism (AFLP)
   DNA analysis   128
Animal and Plant Health Inspection Service
   (APHIS)   40 f.
anthrax   3 ff., 81, 125 ff.
   – antimicrobial therapy   137

   – autoclaving   135
   – bacteriology   125 ff.
   – classification   125 ff.
   – clinical manifestation   130 ff.
   – cutaneous   82, 130 f.
     decontamination   141
   – ELISA   136
   – epidemiology   126
   – gastrointestinal   131
   – genetics   127
   – hemorrhagic meningoencephalitis   134
   – history   125 f.
   – human   130
   – human vaccination   139
   – immunological test   136
   – infant   81
   – infection control   141
   – inhalational   82, 132 ff.
   – intestinal   132
   – investigational therapy   138
   – management   125 ff.
   – microbiological diagnosis   134
   – microbiology   127
   – oral-pharyngeal   132
   – pathogenesis   129
   – PCR   128 ff.
   – post-exposure prophylaxis   137
   – protective antigen (PA)   138
   – serological diagnosis   136
   – toxin receptor (ATR)   129 f.
   – vaccine   140
   – virulence factor   129
antibiotics   86, 192
   – antibiotic susceptibility   47
anticrop agent   4
antimicrobial treatment   50
antitoxin   51, 86
   – therapy   138
antiviral therapy   79, 118

Arenaviridae 197 ff., 208 ff.
  – clinical manifestation 208
  – diagnosis 210
  – disease 208
  – ELISA 210
  – epidemiology 208
  – new world group 207
  – old world group 207
  – PCR 210
  – transmission 210
  – virology 207
Association of Professionals in Infection
  Control and Epidemiology (APIC) 48
autoclaving 24, 49, 135
avian influenza virus 26, 246

b
B cell immunity 130
Bacillus 127
Bacillus anthracis 3 ff., 21, 81, 126 ff.
Bacillus cereus 21
bacteria 10
  – bacterial agent 14 ff.
  – bacterial toxin 130
biodefense facility 35
biohazard containment 48
Biological and Toxin Weapons Convention
  (BWC) 35
biological agent 3
  – category (A, B, C) 78
biological literacy and awareness 37
biological warfare 3
  – program 5
biological weapon 155, 175, 189
  – characteristics 9
  – defensive 4
  – modern 3
  – program 4 f.
  – system 8
Biopreparat 5, 35
biosafety level 48
BioShield Act 234
biosurveillance 46
bioterrorism 1 ff., 26, 251
  – Act 241
  – attack 43
  – bacterial/rickettsial agent 14 ff.
  – biological toxin 23
  – care of children 73 ff.
  – exercise 37
  – feature 156
  – law 244
  – preparedness 33 ff., 53

  – potential 79 ff.
  – potential agent 1 ff., 26
  – threat 26
  – vaccination 80
  – viral agent 22 f.
blood volume 75
blue-green algae 25
Botox 169
botulism 84 ff., 165 ff.
  – antibiotics 86
  – antitoxin 86, 176
  – attack 175
  – biological weapon 174
  – clinical presentation 165 ff.
  – decontamination 86, 178
  – diagnosis 171
  – diagnostic criteria 173
  – differential diagnosis 173
  – ELISA 172
  – epidemiology 166
  – foodborne 84
  – history 165
  – immunization 177
  – infant 85
  – infection control 86, 178
  – isolation of hospitalized children 86
  – management 176
  – microbiology 167
  – PCR 172
  – post-exposure prophylaxis 178
  – prevention 177
  – prognosis 176
  – toxicology 167 ff.
  – transmission 170
  – undetermined etiology 85
  – vaccine 86, 177 f.
  – wound 85
botulinum toxin 4 f., 165
  – select agent 172
  – subtype (A, B, C, D, E, F, G) 165
botulinum toxoid
  – pentavalent botulinum toxoid (ABCDE)
    177
  – vaccine 177 f.
Brucella abortus 14
Brucella canis 14
Brucella melitensis 14
Brucella suis 4 ff.
brucellosis 14 ff.
  – diagnosis 14
  – epidemiology 14
  – management 15
  – microbiology 14

Bunyaviridae 197 ff., 210 ff.
  – ARDS 211
  – clinical manifestation 211
  – diagnosis 213
  – disease 211
  – ELISA 213
  – epidemiology 211
  – PCR 213
  – transmission 213
  – virology 210 f.
Burkholderia mallei 3, 15 ff.
Burkholderia pseudomallei 15

c

Campylobacter jejuni 21
Category A 12 f.
  – agents 149, 183, 194
Category B 12 ff., 23 ff.
Category C 12, 25 ff.
CD4+ 206
CD8+ 206
Centers for Disease Control and Prevention
  (CDC) 5 ff., 50, 128, 172, 194, 231
chemical decontamination 49
Chemical Weapons Convention (CWC) 36
chemoprophylaxis 15 ff.
children 73 ff.
  – developmental factors 75
  – hospitalization 79 ff.
  – increased vulnerability 73 ff.
  – infection control 84
  – isolation precaution 84
  – post-exposure prophylaxis 84
  – vaccine 84
Chlamydophila psittaci 17
cholera 55, 246
civil liberty 250
civilian biodefense 36
clostridial neurotoxin 168
Clostridium botulinum 21, 84, 165 ff.
  – select agent 172
Clostridium perfringens 21 ff.
cognitive ability 75
communication 40
  – adolescent 77
  – children 77
  – electronic information 46
  – system 46
complication
  – gastrointestinal 110
  – neurological 110
  – ophthalmic 110
  – osteo-articular 110

  – respiratory 110
  – skin 109
contagion 9
contamination 139
Council of State and Territorial Epidemiolo-
  gists (CSTE) 37
Coxiella burnetii 4, 18
Crimean-Congo hemorrhagic fever virus
  (CCHF) 211 ff.
Cryptosporidium parvum 21
CSF protein 174

d

decontamination 48 f.
  – chemical 49
  – mechanical 49
  – physical 49
  – shower 76
deer-fly fever 183
delivery system 8
dengue fever virus 214
Department of Health and Human Services
  (DHHS) 140
detection 39
diagnosis
  – delayed in children 76
  – differential 111 ff
diphtheria 246
disease reporting 44
dispersion system 8
disseminated intravascular coagulopathy (DIC)
  153, 187
DNA
  – amplified fragment polymorphism
    (AFLP) DNA analysis 128
  – genome 97
  – target 47
  – virus 96
doxocycline 15 ff.
drug resistance 34

e

eastern equine encephalitis (EEE) virus 22, 215
Ebola 36, 201
  – fatal and non-fatal infection 203
  – virus 5
edema factor (EF) 129 ff.
  – edema toxin 130
electronic
  – Department of Defense Electronic
    Surveillance System for the EARLY
    Notification of Community-Based
    Epidemics (ESSENCE) 46

– information   46
ELISA (enzyme-linked immunosorbent assay)
   136, 172, 189, 210 ff.
emergency
   – emergency department-based emerg-
      ing-infections sentinel network
      (EMERGENCY ID NET)   46
   – response capability   42
   – response contact   61
endoprotease   168
enterobacteriaceae family   150
enterotoxin   23
   – enterotoxin B   23
entomologist   34
epidemic   1
epidemiologist   34
epidemiology   48
   – epidemiologic capability   37
   – Programs in Epidemiology and Public
      Health Interventions Network (TEPHI-
      NET)   54
epsilon (alpha toxin)   24
equipment   76
Escherichia coli   21
evasion of natural or vaccine-induced immunity   39
exfoliative toxin   25
exocytosis of neurotransmitters   168
exotoxin   23

f
F1 antigen   152
failure to be detected   39
FAS/CD95   205
FAS-associated Death Domain (FADD)   205
Federal Emergency Management Agency
   (FEMA)   52
federal public-health authority   243
Filoviridae   197 ff.
   – clinical manifestation   202
   – diagnosis   206
   – disease   202
   – epidemiology   201
   – genetically heterogeneous subtypes
      201
   – pathogenesis   205
   – transmission   204
   – virology   201
first responder   42
Flaviviridae   197 ff., 214 ff.
fluid reserve   75
Food and Drug Administration (FDA, US
   FDA)   50, 169

food safety threat   21
food-testing   47
Foodborne Diseases Active Surveillance Network (FoodNet)   22
foodborne pathogen   47
Francisella philomiragia   184
Francisella tularensis   4, 88, 183 f.
   – pathogenicity island (FPI)   184 f.
   – subspecies   184

g
gastrointestinal (GI) syptoms   14
genetically engineered organsism   34
Geneva protocol   4, 34
Giardia lambia   21
glanders   3, 15 ff.
   – diagnosis   16
   – epidemiology   15
   – management   17
   – microbiology   15
glycoprotein (GP)   201 ff.
   – secreted (sGP)   201
Guarnarito virus   207
Guillain-Barré syndrome (GBS)   173 f.
   – Miller-Fisher variant   173 f.

h
Haemophilus influenzae   189
Hanta virus   211 ff.
Hantaan pulmonary syndrome (HPS)   211
healthcare worker   77
height   74
hemin storage system   152
hemorrhagic fever   197 ff., 246
   – renal syndrome (HFRS)   212
   – virus   5, 197 ff.
hemorrhagic smallpox variola elementary
   bodies (cytoplasmic inclusions)   101
hendra virus   26
hepatitis A   21
   – virus   4
Herpes simplex virus   135
HHS select agents and toxins   60
hospitalized child   79
human botulism immune globulin (HBIG)
   176
human influenza virus   26
hypochlorite solution   49

i
identification   9
   – critical biological agent   47
immune response   114, 140

immunity   9, 39, 51 ff.
incubation period   8 f., 99
infection control   79 ff., 141, 162, 178, 193
infection-control procedure   48
infectivity   9 f.
influenza virus
  – avian   26, 246
  – human   26
  – H5N1   26
Institute for Genomic Research (TIGR)   128
interferon
  – a   205
  – g-producing cell   206
interleukin 1b   205
international biodefense action   34
International Health Regulations (IHR)   55,
  246 f.
isolation   245
  – aeromedical isolation team   48
  – patient isolation precaution   70

j
joint services lightweight integrated suit
  technology (JSLIST)   48
Junin virus   207 ff.

k
Kyasanur forest disease (KFD)   214 f.

l
laboratory
  – criteria for confirmation   112 f.
  – diagnostics   10, 188
  – food-testing   47
  – levels (A, B, C, D)   47
  – microbiology   48
  – preparedness   46 f.
  – response network (LRN)   135, 194
Lambert-Eaton myastenic syndrome (LEMS)
  173
Lassa fever virus   207 ff.
legal preparedness   241 ff.
lethal factor (LF)   130 ff.
  – lethal toxin   130
lipopolysaccharide endotoxin   152
Listeria monocytogenes   21
low-calcium-response (*Lcr*) plasmid   152
Loxoscales recluse   82
lymphocytic choriomengitis virus (LCM)   207

m
Machupo virus   207 ff.
macrophage inflammtory protein (MIP-1a)   205

macrophage-mediated cytokine release   130
Malta fever   14
management   40
Marburg   201 ff.
  – glycoprotein (GP)   201
  – virus   5
mechanical decontamination   49
medication   76
Mediterranean fever   14
melioidosis   15
microbial agents   9, 21
microbiology
  – laboratory   48
mitogen-activated protein kinase kinase
  (MAPKK) pathway   130
Model State Emergency Health Powers Act
  (MSEHPA, Model Act)   39 f., 249 f.
modernization of state   241
monocyte chemotactic protein (MCP-1)
  205
motor skills   75
multi-purpose overboots (MULO)   48
multiple-locus variable-number of tandem
  repeat analysis (VNTR, MLVA)   128, 156
munition   8
Mycoplasma fermentans   139
Myobloc   169

n
National Childhood Vaccine Injury Act
  (NCVIA)   232
National Incident Management System
  (NIMS)   245
National Institutes of Health (NIH)   50, 138
National Laboratory Response Network (LRN)
  47
National Repository of Pharmaceuticals and
  Medical Material (NPS)   52
National Response Plan (NRP)   245
Neisseria meningitidis   4
nematode anticoagulant peptide c2 (NAPc2)
  206
neurotoxins
  – A, B, E, F   84
  – clostridial neurotoxin   168
NF-kB (nuclear factor)   205
Nipah virus   26
norovirus   21

o
Oharas disease   183
Omsk Hemorrhagic fever virus (OHF)   214
outbreak investigation   44

p

Pan American Health Organization (PAHO)
    178
pandemic   1
pandemic influenza   26
passive antibody   51
pathogen
    – drug-resistant   34
    – foodborne   47
pathogenicity island (PI)   184 f.
patient isolation precaution
    – airborne precaution   70
    – contact precaution   70
    – droplet precaution   70
    – standard precaution   70
payload   8
PCR (polymerase chain reaction)   47, 172, 210 ff.
    – LightCycler   47
personal protective equipment   48
phage lysis g (PlyG)   139
phagocytosis   129, 152
pharmaceutical readiness   50
phospholipase D (PLD)   152
physical decontamination   49
plague   10, 55, 87 f., 149 ff.
    – antibiotic prophylaxis   161
    – antibiotic treatment   159
    – biological weapon   155
    – bioterrorismus potential   87, 149 ff.
    – bubonic   153
    – category A agent   149
    – children   87 ff.
    – clinical feature   152
    – diagnosis   156
    – epidemic   149 ff.
    – endemic   149 ff.
    – global epidemiology   150
    – history   149
    – immunization   161
    – infection control   87, 162
    – isolation precaution   87
    – laboratory diagnosis   154
    – microbiology   150
    – mortality   154
    – pathogenesis   151
    – post-exposure prophylaxis   88
    – prevention   160
    – primary pneumonic   153
    – primary septicemic   153
    – prophylaxis of plague   160 f.
    – quarantine   246
    – radiology   155
    – treatment   87, 158 f.

plasmid
    – low-calcium-response (*Lcr*) plasmid
        152
    – pX01   129
    – pX02   129
plasminogen activator   152
Plasmodia spp.   4
'point-source' origin   9
policy priority   227 ff.
    – smallpox   227
    – stockpile   227
    – surveillance   227
political preparedness   52
poly g-D-glutamic acid capsule (gDPGA)   129
polymerase chain reaction, *see* PCR
post traumatic stress disorder (PTSD)   75
potential bioterrorism agent   10 ff.
    – categorization   10
    – prioritization   10
Poxviridae   96 f.
    – *Chordopoxviridae*   97
    – *Orthopoxvirus*   97
preparedness   40, 53
    – political   52
prevention   140
priority category (A, B, C)   11
proapoptotic gene   205
prophylaxis   50, 161
protection   5, 40
protective antigen (PA)   129 ff.
Pseudomonas aeruginosa   140
    – exotoxin A (rEPA)   140
psittacosis   17
    – diagnosis   17
    – epidemiology   17
    – management   18
psychological injury   75
public health   34
    – emergency   249
    – law   40, 241 ff.
    – legal preparedness   248
    – Public Health Security and Bioterrorism
        Preparedness and Response Act   241
    – system   194
    – system preparedness   42

q

Q fever (query fever)   18
    – diagnosis   19
    – epidemiology   18
    – management   19
    – microbiology   18
quarantine power   245 f.

r

rabbit fever   183
rash of smallpox
    – centrifugal spread   103 f.
recognition   42
reporting   194
resistance to antimicrobial agent   39
respiratory distress syndrome
    – adult (ARDS)   153, 187, 211
respiratory rate   74
ribavirin   209 ff.
ricin toxin   24
rickettsia   10 ff.
Rickettsia maoseri   4
Rickettsia prowazekii   4, 19
rifampin   15
Rift Valley fever virus (RVF)   211 ff.
RNA
    – virus   197

s

Sabia virus   207
Salmonella sp.   21
Salmonella thypi   21
SARS, *see* severe acute respiratory syndrome
sausage poison   165
saxitoxin   25
select agent   35, 60, 172
septic shock   130
Serratia marcescens   5
serum agglutinations test   15
severe acute respiratory syndrome (SARS)   2,
    27, 246
    – SARS-associated coronavirus (SARS-
    COV)   27
Shigella spp.   4, 21
Sin Nombre virus   211
single nucleotide polymorphism (SNP)   129
skin   74 f.
smallpox (variola)   10, 78, 95 ff.
    – antiviral therapy   79, 118
    – bioterrorism potential   79
    – classification   98 f.
    – clinical presentation   95 ff.
    – complication   109
    – death   105
    – differential diagnosis   111
    – early hemorrhagic-type   108
    – Emergency Personnel Protection Act
    (SEPPA)   234
    – eruptive stage   102
    – febrile prodrome   113

    – flat-type   106
    – hemorrhagic-type   107
    – history   96
    – immunity   116
    – incubation period   99
    – individual lesion   102
    – infection control   79
    – isolation of hospitalized child   79
    – laboratory diagnosis   115
    – late hemorrhagic-type   108 f.
    – lesion   113
    – modified-type   106
    – mortality rate   107
    – ordinary type   102
    – pathophysiology   113
    – pre-eruptive stage   100
    – pre-event vaccination   227 ff.
    – preparedness   227
    – presentation   95 ff.
    – prognosis   98
    – postexposure infection control   115
    – quarantine   246
    – rash   98 ff.
    – stages   98
    – symptom   100 f.
    – vaccine   80, 116, 227
    – virology   95 ff.
    – virus   5
SNAP-25   168
SNARE (soluble *N*-ethylmaleimide-sensitive
    factor attachment protein receptor)   168 f.
Staphylococcus aureus   21 ff.
Staphylococcus enterotoxin B (SEB)   4, 23
stockpiling   52, 227 ff.
    – medical therapeutic   11
surface-to-mass ratio   74
susceptibility
    – antibiotic   47
surveillance   10, 40 ff., 227 ff.
syntaxin   168

t

T cell immunity   130
T-2 mycotoxin   24
Tacribe virus   207
tetanus   25
tetracycline   17
terodotoxin   25
TF-specific hookworm inhibitor nematode
    anticoagulant peptide c2 (NAPc2)   206
The Hague Convention   34
TNF, *see* tumor necrosis factor

TNF-related apoptosis-inducing ligand (TRAIL)/Apo2L   205
TOPOFF (top officials)   37 f., 249
toxic-shock syndrome toxin (TSST-1)   25
toxicity   10
toxin   5 ff.
– anthrax   138
– antitoxin   51
– bacterial   130
– biological   23
– botulinum   4, 165
– category (A, B, C)   78
– clostridial neurotoxin   168
– edema   130
– exfoliative   25
– exotoxin A (rEPA)   140
– human botulism immune globulin (HBIG)   176
– lethal   130
– lipopolysaccharide endotoxin   152
– neurotoxin   84, 168
– ricin   24
– *Staphylococcus* enterotoxin B (SEB)   4, 23
– toxin-induced lysis of infected macrophage   130
– tripartite   129
training   42, 77
– Training Programs in Epidemiology and Public Health Interventions Network (TEPHINET)   54
transmission   39
– person-to-person   11
tripartite toxin   129
tularemia   88 ff., 183 ff.
– antibiotics   192
– antimicrobial agent   191
– biological weapon   189
– bioterrorismus potential   89
– category A agent   183
– children   88 ff.
– classification   184
– clinical forms   186
– diagnostic criteria   190 f.
– ELISA   189
– epidemiology   185
– glandular   187
– history   183
– immunization   193
– infection control   89, 193
– isolation precaution   89
– laboratory diagnosis   188
– microbiology   184

– natural disease   183 ff.
– oculoglandular   187
– oropharyngeal   187
– pathogenesis   186
– PCR   189
– pneumonic   187
– post-exposure prophylaxis   89, 193
– prevention   193
– radiology   189
– septicemia   187
– treatment   90, 191
– typhoidal   187
– ulceroglandular   187
– vaccine   90
– Western blot (Wb)   189
tumor necrosis factor (TNF)   205
typhus fever   19 ff.
– diagnosis   20
– management   20

u

undulant fever   14
US Army Medical Research Institute for Infectious Disease (USAMRIID)   5, 47, 128 ff.
USDA select agents and toxins   60
UV   49

v

V antigen   152
vaccination   80, 139, 227
– pre-event vaccination   227 ff.
vaccine   51 ff.
– absorbed anthrax (AVA)   139
– anthrax   140
– botulinum toxoid   177
– dually active anthrax (DAAV)   141
– induction of false positivity   4
– tetravalent   178
vaccinia
– immune globulin (VIG)   118
– virus   135
VAMP/synaptobrevin   168 f.
varicella
– rash   79
– Varicella-zoster virus   135
variola   78, 97 f., 114
– sine eruptione   106
– subclinical infection   106
vascular permeability   205
vendor-manged inventory (VMI)   52
Venezuelan equine encephalitis (VEE) virus   4, 22, 178, 215
veterinary medicine   34

Vibrio cholerae 4, 21, 140
Vibrio parahemolyticus 21
viral hemorrhagic fever 197 ff., 246
    – bioterrorismus 197
    – differentiation 197
    – natural disease 197
    – virus 197
virus 10
    – alphavirus 22 f., 215
    – *Arenaviridae* 197 ff., 208 ff.
    – *Bunyaviridae* 197 ff., 210 ff.
    – Crimean-Congo hemorrhagic fever
      virus (CCHF) 211 ff.
    – dengue fever virus 214
    – DNA virus 96 f.
    – Eastern equine encephalitis (EEE) virus
      22, 215
    – Ebola virus 5, 201 ff.
    – *Filoviridae* 197 ff.
    – *Flaviviridae* 197 ff.
    – Guarnarito virus 207
    – Hanta virus 211 ff.
    – hemorrhagic fever virus (HVF) 5,
      197 ff.
    – hendra virus 26
    – hepatitis A virus 4
    – Herpes simplex virus 135
    – influenza virus 26, 246
    – Junin virus 207 ff.
    – Lassa fever virus 207 ff.
    – lymphocytic choriomengitis virus
      (LCM) 207
    – Machupo virus 207 ff.
    – Marburg virus 5, 201 ff.
    – Nipah virus 26
    – norovirus 21
    – Omsk Hemorrhagic fever virus (OHF)
      214
    – Rift Valley fever virus (RVF) 211 ff.

    – Sabia virus 207
    – Sin Nombre virus 211
    – smallpox virus 5
    – Tacribe virus 207
    – vaccinia virus 135
    – Varicella-zoster virus 135
    – Venezuelan equine encephalitis virus
      (VEE) 4, 22, 178, 215
    – viral agent 22 f.
    – western equine encephalitis (WEE)
      virus 22, 215
VP24 (membrane protein) 201
VP30 (minor nucleocapsid) 201
VP40 (matrix) 201
vulnerability 73 ff.

w
W antigen 152
war 2
water safety threat 21
weaponization 9 f.
Weil Felix test 4, 20
Western blot (Wb) 189
western equine encephalitis (WEE) virus 22,
    215
World Health Organization (WHO) 36, 53 f.,
    78, 98, 127, 150, 246 f.
    – biological weapons list 11
Wurst-Kerner 165

x
Xenopsylla cheopis 149

y
yellow fever 55, 214, 246
Yersinia pestis 4, 87, 149
    – virulence factor 152
Yersinia speudotuberculosis 150
Yersinia outer proteins (Yops) 152